Foundations of Materials Science and Engineering

Foundations of Materials Science and Engineering

Julie Barker

WILLFORD PRESS
www.willfordpress.com

Published by Willford Press,
118-35 Queens Blvd., Suite 400,
Forest Hills, NY 11375, USA

ISBN: 978-1-64728-496-1

Cataloging-in-Publication Data

Foundations of materials science and engineering / Julie Barker.
p. cm.
Includes bibliographical references and index.
ISBN 978-1-64728-496-1
1. Materials. 2. Materials science. I. Barker, Julie.
TA403 .F86 2023
620.11--dc23

For information on all Willford Press publications
visit our website at www.willfordpress.com

WILLFORD PRESS

Contents

Preface

Materials science refers to a unique combination of science and engineering, chemistry and physics, and imagination and logic. It is concerned with the discovery and design of novel materials, particularly solids. The four main types of materials studied and used in materials science include polymers, ceramics, metals, and semiconductors. Materials science and engineering (MSE) refers to an interdisciplinary study of useful matter. It applies principles of chemistry, physics, and engineering to solve practical issues related to biotechnology, manufacturing, nanotechnology, energy, and information technology. This field is concerned with material behavior with the goal of comprehending, enhancing and even creating new materials. MSE is utilized in a variety of industries such as transportation, construction, energy, drug delivery, nanotechnology, and tissue engineering. This book is compiled in such a manner, that it will provide an in-depth knowledge about materials science and engineering. It is appropriate for student seeking detailed information in this area as well as for experts.

This book is the end result of constructive efforts and intensive research done by experts in this field. The aim of this book is to enlighten the readers with recent information in this area of research. The information provided in this profound book would serve as a valuable reference to students and researchers in this field.

At the end, I would like to thank all the authors for devoting their precious time and providing their valuable contribution to this book. I would also like to express my gratitude to my fellow colleagues who encouraged me throughout the process.

Julie Barker

1

Classification and Properties of Engineering Materials

1.1 Classification

Ferrous Materials

In all fields, iron is an indispensable material. The materials containing the base metal iron as iron are called ferrous materials.

Ferrous materials can be classified as:

- Pig Iron.
- Wrought Iron.
- Cast Iron.
- Steels.

The following are ferrous metals and the kind of uses to which they are usually put:

- Mild Steel: Carbon content of 0.1 to 0.3% and Iron content of 99.7 – 99.9%. Used for engineering purposes and in general, none specialised metal products.
- Carbon steel: Carbon content of 0.6 to 1.4% and Iron content of 98.6 to 99.4 %. Used to make cutting tools such as drill bits.
- Stainless Steel: Made up of Iron, nickel and chromium. It resists staining and corrosion and is therefore used for the likes of cutlery and surgical instrumentation. Our info graphic celebrating 100 years of stainless steel usage in buildings or the different types of stainless steel.
- Cast Iron: Iron at 94 to 98% and Carbon 2 – 6%. Very strong but brittle. Used to manufacture items such as engine blocks and manhole covers.
- Wrought Iron: Composed of almost 100% of iron. Used to make items such as ornamental gates and fencing.

Manufacture of Steel

Steel is the most widely used of all the metals. It is an alloy of iron and carbon. The carbon content varies between 0.2 to 1.7 % for various grades of steel. Carbon gives the property of hardening ability to steel. Other elements like nickel, thromium, vanadium and tungsten are added to produce specific properties such as ease of machining, toughness and the ability to resist wear or corrosion.

Steel can be manufactured by number of processes are:

- Cementation Process.
- Crucible Process.
- Bessemer Process.
- Open hearth Process.
- L.D Process.
- Duplex Process.
- Electric Process.

Aluminium (A_1)

- Acts as deoxidizer.
- Produces fine austenitic grain size.
- It promotes nit riding.

Chromium (C_r)

Joins with carbon to form chromium carbide, thus adds to depth hardenability with improved resistance the abrasion and wear:

- Increases corrosion and oxidation resistance.
- Increases hardenability.
- Increases high temperature strength.
- Resists abrasion and wear (with high carbon).

Uses:

- 1.0-1.5% Cr in medium and high carbon steels for gears, axles, shafts, springs, ball bearings and metal working rolls.
- 12-30% Cr in martensitic and ferritic stainless steels.

Nickel (Ni)

- Increases toughness and resistance to impact.
- Lessens distortion in quenching.
- Lowers the critical temperature of steel and widens the range of successful heat treatment.
- Strengthens Steels.
- Renders high-chromium iron alloys austenitic.
- Does not unite with carbon.

Molybdenum (Mo)

- Improves high temperature creep resistance.
- Reduces temper brittleness in Ni-Cr steels.
- Stabilizes carbides.
- Increases hardenability.
- A constituent of high-speed tool steels and creep resistant steels.
- Up to 0.5% Mo often added to pearlitic Ni-Cr steels to reduce temper-brittleness.
- Promotes hardenability of steel.
- Makes steel fine grained.
- Makes steel unusually tough at various hardness levels.
- Counter acts tendency towards temper brittleness.
- Raises tensile and creep strength at high temperature.
- Enhances corrosion resistance in stainless steels.
- Forms abrasion resisting particles.

Silicon (Si)

- Improves oxidation resistance.
- Strengthens low alloy steels.
- Acts as a deoxidizer.

Manganese (Mn)

- Contributes markedly to strength and hardness (but to a lesser degree than carbon).
- Counter acts brittleness from sulphur.
- Lowers both ductility and weld ability if it is present in high percentage with high carbon content in steel.

Vanadium (V)

- It is powerful carbide former.
- Stabilized cementite and improves the structure of the chill.

Copper (Cu)

- Increases resistance at atmospheric corrosion.
- Acts as a strengthening agent.

1.2 Engineering Properties of Materials

Engineering Materials

Materials play an important role in the construction and manufacturing of various parts and components. An appropriate selection of a material for a given application adds to economy, working and life of the final part and component.

Properties of Engineering Materials

Material property is the identity of material, which describes its state and behavior under different conditions. The material properties can be broadly categorized as chemical, physical, mechanical and thermal.

The physical properties define the physical state of the material and are independent of its chemical nature. The physical properties of the engineering materials include appearance, mass, texture, density, boiling point, Melting point, viscosity, etc. The chemical properties describe the reactivity of the material and are always mentioned in terms of the rate at which the material changes its chemical identity, e.g. oxidation rate, corrosion rate, etc. The mechanical properties describe the resistance against deformation, in particular, under the static and dynamic mechanical loading condition. The mechanical properties include Poisson's ratio, elastic modulus, yield strength and ultimate tensile strength, hardness and toughness, etc. The thermal properties describes the material behavior under thermal loading and include specific heat, thermal conductivity, thermal diffusivity, coefficient of thermal expansion, etc. For given application or service, an engineering material is selected based on a set of appropriate material properties, often referred to as attributes that would be requisite to sustain various expected loads. Figure depicts a schematic representation of material family, which is utilized in selection of materials for a target application.

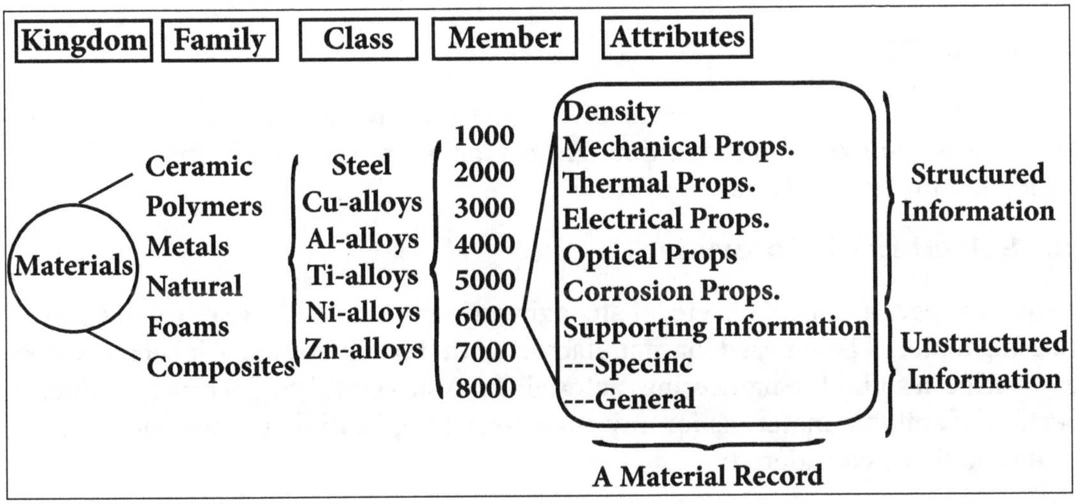

Organized classification of materials and properties.

Physical Properties

Physical properties describe the state of material, which is observable or measurable. Color, density, texture, boiling point, melting point, etc. are some of the commonly known physical properties.

- Color: Represents reflective properties of substance.
- Density: Amount of mass contained by unit volume of material. The higher the density the heavier is the substance, (SI unit: kg/m3).
- Melting point: Melting point is the temperature at which material changes its state from solid to liquid, (SI units: K).
- Boiling point: Boiling point is the temperature at which material changes its state from liquid to gaseous, (SI units: K).

Chemical Properties

Chemical properties are the measure of reactivity of a material in the presence of another substance or environment which imposes a change in the material composition. These properties are always mentioned in term of rate of change in its composition. Oxidation rate, Corrosion rate, etc. are some of the chemical properties of material.

Oxidation Rate

Oxidation rate is measured in terms of the amount of material consumed forming oxide or amount of oxide scale formed for a given period of time at specific surrounding temperature. Oxidation rate is given by the amount of mass of material lost or thickness of scale formed during the oxidation per unit time. (Units: gms/min or µm/min).

Corrosion Rate

Corrosion rate is measured in terms of the corrosion penetration for given period of time at specific surrounding condition. The corrosion rate is given by the length of penetration per unit time. (Units: mm/year).

1.2.1 Selection of Materials

The quality, performance, life etc. of an engineering product, all are effected by the engineering material being used for manufacturing that product. Hence it becomes necessary to select a suitable Engineering materials for a successful engineering product. For selection of suitable materials for any engineering application and product, following factors should be considered:

- Mechanical strength.
- Stability.
- Ductility.
- Availability.
- Fabric ability.
- Design.
- Corrosion Resistance.
- Cost.

When we talk about choosing the materials for a component, we take into account many different factors. These factors can be broken down into the following areas;

Material Properties: The expected level of performance from the material Cost and Availability

Material must be priced appropriately (not cheap but right). Material must be available.

Processing: Must consider how to make the part, for example: Casting Machining Welding.

Environment

- The effect that the service environment has on the part.
- The effect the part has on the environment.
- The effect that processing has on the environment.

Now clearly these issues are inter-linked in some of the fashion. For example, the cost is a direct result of how difficult the material is to obtain and to machine. And the effect of the environment on the material is clearly related to its material properties.

So if we really want to use a novel or unusual material, the choice must be made early in the design process. Then we can do the detailed design work using the correct material properties.

Consider the example of wooden airplanes and metal-framed airplanes. If we were to design an airplane of either material, we will have to make the choice early. The end designs are quite different. So, the material choice can radically alter the final design.

1.3 Mechanical Properties of Materials: Tensile Strength

The stress induced at any cross section of the member which is subjected to a tensile force is referred as tensile stress,

$$\text{Tensile} = \left(\frac{\text{Tensile Force}}{\text{Area of cross section}}\right)$$

1.3.1 Stress: Strain Behavior

Stress strain curve is a behavior of material when it is subjected to load. In this diagram stresses are plotted along the vertical axis and as a result of these stresses, corresponding strains are plotted along the horizontal axis. As shown below in the stress strain curve.

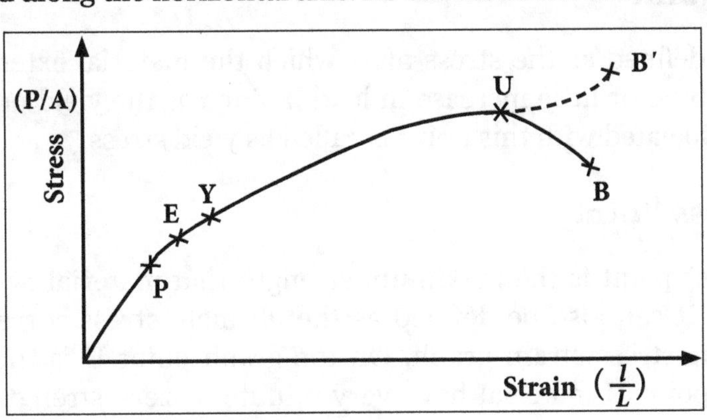

Stress: strain curve.

From the diagram one can see the different mark points on the curve. It is because, when a ductile material like mild steel is subjected to tensile test, then it passes various stages before fracture.

These stages are:

- The Proportional Limit.
- The Elastic Limit.
- The Yield Point.
- The Ultimate Stress Point.
- The Breaking Point.

Proportional Limit

Proportional limit is a point on the curve up to which the value of stress and strain remains proportional. From the diagram point P is the referred the proportional limit point or it can also be known as the limit of proportionality. The stress up to this point can be also be referred as the proportional limit stress.

Hook's law of proportionality from the diagram can be defined between the point OP. It is so, because the OP is a straight line which shows that Hook's law of stress strain is followed up to point P.

Elastic Limit

Elastic limit is the limiting value of stress up to which the material is in perfectly elastic condition. From the curve, the point E is the elastic limit point. Material will return back to its original position, if it is unloaded before the crossing of point E. This is so, because the material is perfectly elastic up to point E.

Yield Stress Point

A yield stress is defined as the stress after which the material extension takes place more quickly with no or little increase in load. Point Y is the yield point on the graph and the stress associated with this point is called as yield stress.

Ultimate Stress Point

An ultimate stress point is the maximum strength that material have to bear stress before breaking. It can also be defined as the ultimate stress corresponding to the peak point on the stress strain graph. On the graph point U is the ultimate stress point. After the point U material have very minute or zero strength to face further stress.

Breaking Stress (Point of Rupture)

Breaking point or the breaking stress is point where the strength of material breaks. The stress associates with this point is referred as breaking strength or rupture strength. On the stress strain curve, the point B is the breaking stress point.

1.3.2 Ductile and Brittle Material

The pendulum swings through during the test, the height of the swing being a measure of the amount of energy absorbed in fracturing the specimen. Conventionally, three specimens are tested at any one temperature and the results averaged.

Charpy tests show whether a metal can be classified as being either ductile or brittle. This is particularly useful for the ferritic steels that show a ductile to brittle transition with a decreasing temperature. A brittle metal will absorb a small amount of energy when impact tested, a tough ductile metal absorbs a large amount of energy. The appearance of the fracture surface also gives information about the type of fracture that has occurred; a brittle fracture is bright and crystalline, a ductile fracture is dull and fibrous. The percentage crystallinity is determined by making a judgement of the amount of crystalline or brittle fracture on the surface of the broken specimen and is a measure of the amount of brittle fracture.

Lateral expansion is a measure of the ductility of the specimen. When a ductile metal is broken, the test-piece deforms before breaking and the material is squeezed out on the sides of the compression face. The amount by which the specimen deforms in this way is measured and expressed as mille metres of lateral expansion.

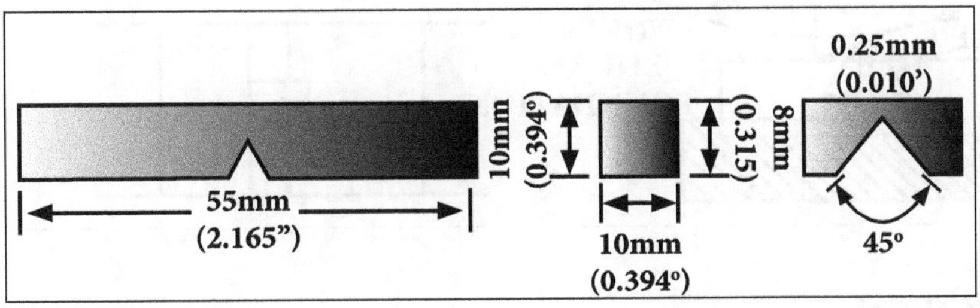

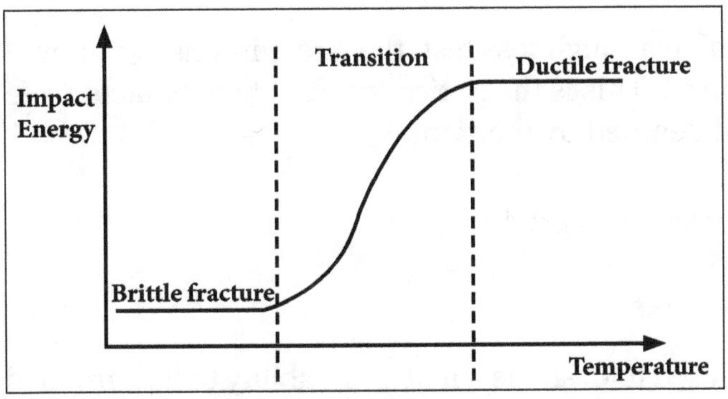

Ductile to Brittle transition curve.

When reporting the results of Charpy test, the absorbed energy (in J) is always reported, while the percentage crystallinity and lateral expansion are optional on the test report. It should be emphasised that the Charpy tests are qualitative, the results can only be compared with each other or with the requirement in a specification. They can not be used to calculate the fracture toughness of a weld or parent metal.

1.3.3 Impact Test

In the Izod impact test, the test piece is the cantilever that is clamped upright in an anvil, with a V notch at the level of the top of the clamp. The test piece is hit by a striker carried on a pendulum which is allowed to fall freely from a fixed height, to give a blow of 120 ft lb energy. After fracturing the test piece, a height to which the pendulum rises is recorded by a slave friction pointer mounted on the dial, from which the absorbed energy amount is read.

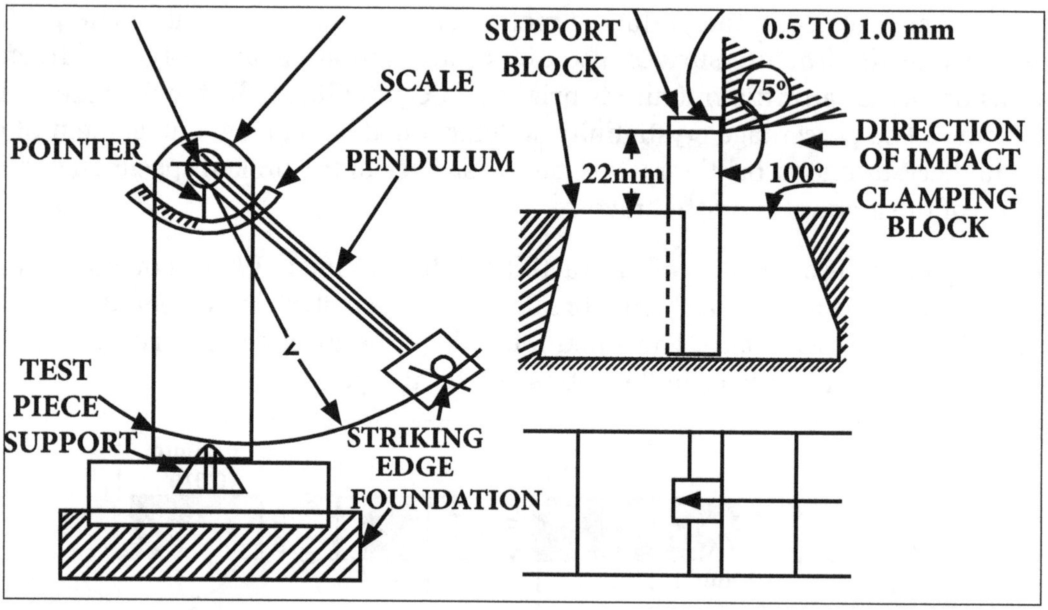

1.3.4 Toughness

For a given material in a toughness test, the stress intensity factor at which a crack will propagate is measured. This is the critical stress intensity factor referred to as the fracture toughness and denoted by the symbol,

$$N=\frac{K_{IC}}{K_I}(N=\text{factor of safety})$$

1.3.5 Hardness Test

Hardness is a term used to describe a material's ability to be hardened when it is exposed to heat and quenched or cooled rapidly.

It should not be confused with hardness which refers to materials' strength and ability to resist damage

Hardness Testing

Hardness may be defined as resistance of metal to plastic deformation (usually) by indentation. However, the term may also refer to stiffness or temper or to resistance to scratching, abrasion or cutting.

Indentation hardness may be measured by various hardness tests such as Brinell, Rockwell etc.

1.3.6 Fatigue and Fatigue Test

Fatigue is defined as the capacity of material to withstand repeatedly applied stresses

The resistance of a material to fatigue failure is characterized by its fatigue or endurance limit.

The endurance limit or endurance strength is defined as the maximum stress which a specimen can endure without failure when this stress is repeated for a specified number of cycles.

Fatigue tests determine the resistance of material to repeated pulsating or fluctuating loads.

1.3.7 Creep and Creep Test, Fracture

This method is used for determining creep or stress relaxation behavior. To determine creep properties, a material is subjected to prolonged constant tension or compression loading at constant temperature.

When a material sample is subjected to constant stress, an increase in the deformation takes place with time, this behavior is known creep. The processes causing the creep may be physical or chemical in nature and under all normal conditions both types of process will occur simultaneously.

However at normal or at low temperatures, creep is dominated by physical processes. At high temperatures or long times, the chemical processes are dominant. One type of creep instrument measures the straining in the test specimen continuously and consists of a small rig with a chamber and a clip on extensometer.

Dead weights are placed on a loading pan with a lever arm to increase the effect of the weight by a factor of 10. The rigs are connected to a data acquisition box connected to a computer, storing the strain data over time.

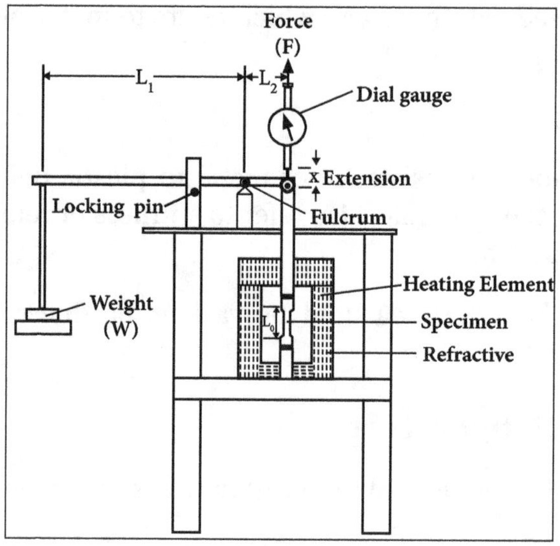

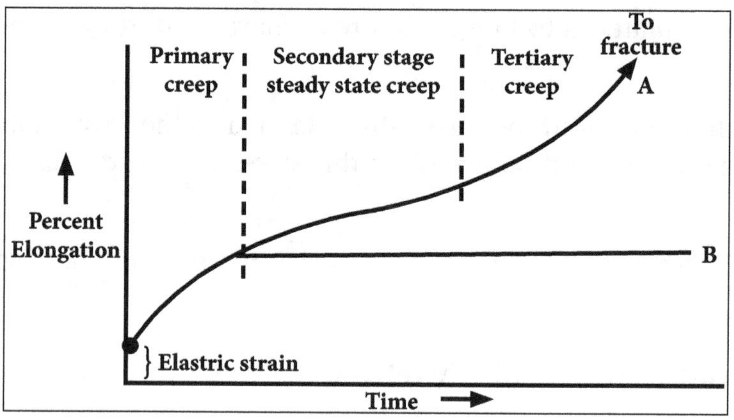

Creep curve.

Fracture

Fracture is an inhomogeneous process of deformation that makes regions of material to separate and load-carrying capacity to decrease to zero. It can be viewed on many levels, depending on the size of the fractured region that is of interest. At the atomistic level fracture occurs over regions whose dimensions are of the order of the atomic spacing (i.e., 10-7 mm); at the microscopic level fracture occurs over regions whose dimensions are of the order of the grain size (of about 10-3 mm); and at the macroscopic level fracture occurs over dimensions that are of the order of the size of flaws or notches (1 mm or greater).

2

Electrical and Electronic Materials

2.1 Electrical and Thermal Conductivity

Electrical conductivity:

The electrical conductivity is defined as the quantity of electricity flowing per unit area per unit time at a constant potential gradient,

$$\sigma = \frac{ne^2\tau}{m} \text{ Ohm}^{-1}\text{m}^{-1}$$

Expression for Electrical Conductivity

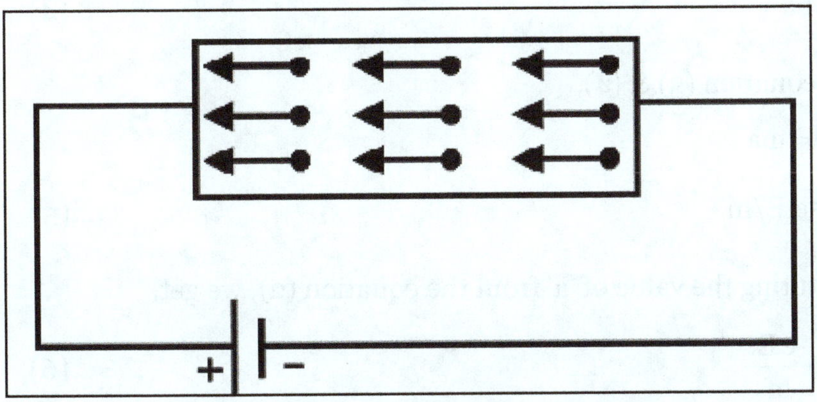

Moment of Electron.

When an electric field (E) is applied to a conductor, the free electrons are accelerated and give rise to current (I) which flows in the direction of electric flows of charges is given in terms of current density.

Let 'n' be the number of electrons per unit volume and 'e' be the charge of the electrons.

The current flowing through the conductor per unit area in unit time (current density) is given by,

$$J = nV_d\ (-e)$$

$$J = -nV_d\ (e) \qquad \text{...(1)}$$

The negative sign indicates that the direction of current is in opposite direction to the movement of electron.

Due to the applied electric field, the electrons acquire an acceleration 'a' can be given by,

$$\text{Accelaration }(a)=\frac{\text{Drift Velocity }(V_d)}{\text{Relaxation time}(\tau)}$$

$$a=V_d/\tau$$

$$V_d=a\tau \qquad ...(2)$$

When an electric field of strength (E) is applied to the conductor, the force experienced by the free electrons is given by,

$$F=-eE \qquad ...(3)$$

From Newton's second Law of motion, the force acquired by the electrons can be written as,

$$F=ma \qquad ...(4)$$

Comparing equation (3) & (4),

$$-eE=ma$$

$$a=-e\,E/m \qquad ...(5)$$

Now, substituting the value of 'a' from the equation (2), we get,

$$V_d=\frac{-e\,E\tau}{m} \qquad ...(6)$$

Substitute equation (6) in (1), we get,

$$J=n\left(\frac{-e\,E\tau}{m}\right)(-e)$$

$$J=\frac{ne^2\,E\tau}{m} \qquad ...(7)$$

The electrical conductivity, $\sigma=J/E$

The electrical conductivity is, $\sigma=\frac{ne^2\tau}{m}$

Where,

m - Effective mass of free electron.

σ - Electrical conductivity.

τ - Relaxation time.

n - Number of electrons.

Thermal Conductivity (K)

The thermal conductivity is defined as the amount of heat flowing through an unit area per unit temperature gradient,

$$K = -\frac{Q}{A\left[\frac{dT}{dx}\right]} \ Wm^{-1}K^{-1}$$

The negative sign indicates that heat flows from hot end to cold end.

Where,

K is the thermal conductivity of metal.

Q is the amount of heat energy.

dT /dx is the temperature gradient.

In general, the thermal conductivity of a material is due to the presence of lattice vibrations (i.e., photons and electrons). Hence, the total thermal conduction can be written as,

$$K_{total} = K_{electron} + K_{photons}$$

Expression for thermal conductivity (K) of an electron.

Consider a metal bar with two planes A and B separated by a distance 'λ' from C. Here T_1 is hot end and T_2 is cold end. i.e., $T_1 > T_2$.

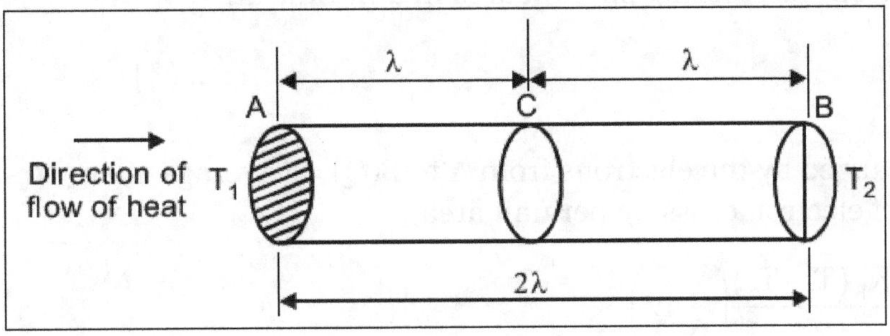

Thermal Conductivity.

Let,

'n' be the number of conduction electrons and 'v' be the velocity of the electrons.

K_B is the Boltzmann constant.

From kinetic theory of gases,

Energy of an electron at $A = \frac{1}{2}mv^2$

$$= \frac{3K_B T_1}{2} \quad ...(1)$$

The kinetic energy of an electron at $B = \frac{1}{2} mv^2$

$$= \frac{3K_B T_2}{2} \quad ...(2)$$

The net energy transferred from A to B

$$= \frac{3K_B (T_1 - T_2)}{2} \quad ...(3)$$

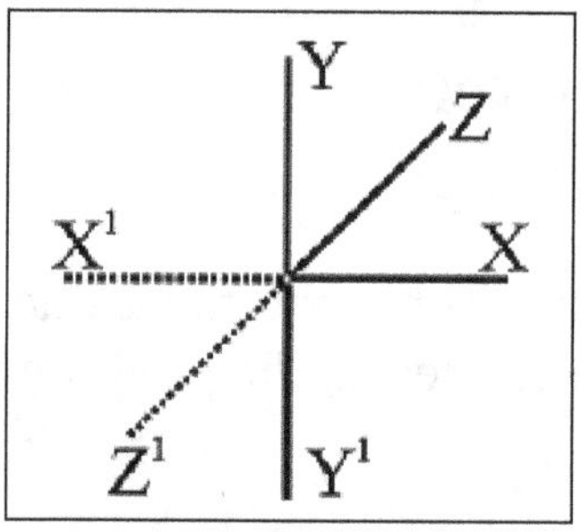

Moment of electron field.

Let us assume that there is equal probability for the electrons to move in all the six directions. Each electrons travels with thermal velocity 'V' and 'n' is the free electron density then an average of 1/6 nv electron will travel in any one direction.

Number of electrons crossing per unit area in unit time at,

$$C = \frac{1}{6}nv \quad ...(4)$$

The energy carried by the electrons from A to B(Q) = (Average energy transfer from A to B) * (No. of electron crossing per unit area),

$$Q = \frac{3K_B (T_1 - T_2)}{2}\left(\frac{1}{6}nv\right)$$

$$Q = \frac{1}{4} K_B (T_1 - T_2) nv \qquad ...(5)$$

We know that the thermal conductivity, $K = \frac{Q}{A\left(\frac{dT}{dx}\right)}$.

The heat energy transferred per unit sec per unit area is, $Q = K\frac{dT}{dx}$ [A = 1 unit area],

$$Q = \frac{K(T_1 - T_2)}{2\lambda} \qquad ...(6)$$

$$dT = T_1 - T_2,\ dx = 2\lambda$$

Comparing equations (5) and (6),

$$\frac{K(T_1 - T_2)}{2\lambda} = \frac{1}{4} K_B (T_1 - T_2) nv$$

Thermal conductivity is given by,

$$K = \frac{1}{2} K_B nv\lambda$$

2.1.1 Free Electron Theory

It was developed by Drude and Lorentz and hence is also known as Drude-Lorentz theory. According to this theory, a metal consists of electrons which are free to move in the crystal like molecules of the gas in the container. Mutual repulsion between electrons is ignored and hence potential energy is taken as zero. Therefore the total energy of the electron is equal to its kinetic energy.

Postulates of Classical Free Electron Theory

- A solid metal has nucleus with revolving electrons. The electrons move freely like molecules in a gas.
- The free electrons move in a uniform potential field due to the ions fixed in the lattice.
- In the absence of electric field ($E = 0$), the free electrons move in random directions and collide with each other. During this collision no loss of energy is observed, since the collisions are elastic as shown in figure.
- In the presence of electric field ($E \neq 0$), the free electrons are accelerated in the direction opposite to the direction of applied electric field, as shown in figure.

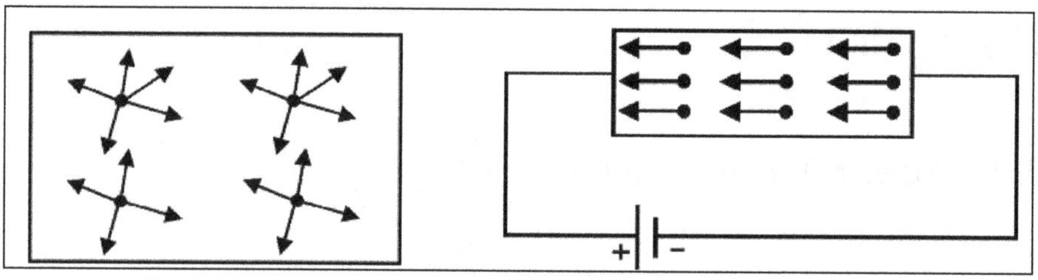

Absence of electric field (E = o) & Presence of electric field (E ≠ o).

- Since the electrons are assumed to be perfect gas, they obey the laws of classical theory of gases.
- Classical free electrons in the metal obey Maxwell-Boltzmann statistics.

Uses of Classical Free Electron Theory

- It is used to verify Ohm's law.
- It is used to explain electrical conductivity (σ) and thermal conductivity of (K) of metals.
- It is used to derive Wiedemann-Franz law.
- It is used to explain the optical properties of metal.

Drawbacks of Classical Theory

- It is a macroscopic theory.
- According to classical free electron theory, all the free electrons will absorb energy but the quantum free electron theory states that only few electrons will absorb energy.
- This theory cannot explain the Compton effect, Photo-electric effect, para-magnetism and ferromagnetism etc.
- This theory cannot explain the electrical conductivity of semiconductors and insulators.
- Dual nature of light radiation cannot be explained.
- The theoretical and experimental values of specific heat and electronic specific heat are not matched.
- By classical theory $K/\sigma = T$ is constant for all temperature but by quantum theory $K/\sigma = T$ is not a constant for all temperatures.
- The Lorentz number obtained by classical theory does not have good agreement with experimental value and theoretical value it is rectified by quantum theory.

Conductors are those materials in which plenty of free electrons are available for electric conduction.

In terms of energy bands, it means that electrical conductors are those which have overlapping valence and conduction bands as shown in figure. In fact, there is no physical distinction between the two bands, hence the availability of a large number of conduction electrons.

Hence the free electrons can easily move from the valence band to conduction band and are available for electrical conduction under the action of an electric field.

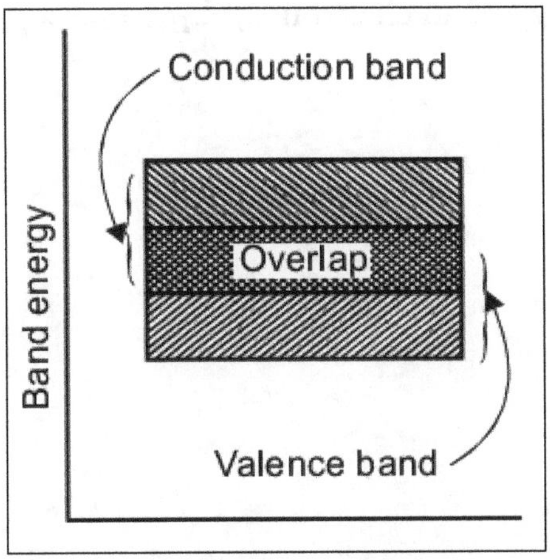

Energy band gap – conductor.

Electrical Conductivity - Wiedemann-Franz Law

The ratio between the thermal conductivity (K) and electrical conductivity (σ) of a metal is directly proportional to the absolute temperature of the metal,

$$\frac{K}{\sigma} \propto T \quad \text{or} \quad \frac{K}{\sigma} = LT$$

Where L is called Lorentz number, the value of L is 2.44 × 10–8WΩK–2(as per Quantum Mechanical value).

Derivation

By Classical theory, we can derive Wiedemann-Franz law using the expressions for electrical and thermal conductivity of metals.

The expression for thermal conductivity, $K = \frac{K_B n v \lambda}{2}$

The expression for electrical conductivity, $\sigma = \frac{ne^2 \tau}{m}$

$$\frac{K}{\sigma}=\frac{1/2\, K_B\, nv\lambda}{ne^2\, \tau/m}$$

$$\frac{K}{\sigma}=\frac{1}{2}\frac{mK_B\, v^2}{e^2} \qquad \left(\because \frac{\lambda}{\tau}=v\right)$$

$$\frac{K}{\sigma}=\frac{1}{2}mv^2\frac{K_B}{e^2}$$

We know that kinetic energy of an electron is, $\frac{1}{2}mV^2=\frac{3}{2}K_BT$,

$$\frac{K}{\sigma}=\frac{3}{2}K_B\, T\frac{K_B}{e^2}$$

$$\frac{K}{\sigma}=\frac{3}{2}\frac{K_B{}^2T}{e^2}$$

$$\frac{K}{\sigma T}=\frac{3}{2}\frac{K_B{}^2}{e^2}$$

$$\frac{K}{\sigma T}=L$$

L is called Lorentz number.

Thus, it is proved that the ratio of thermal conductivity and electrical conductivity of a metal is directly proportional to the absolute temperature of the metal.

Lorentz Number: It is the ratio of thermal conductivity of a metal to the product of its electrical conductivity and its absolute temperature.

Thus,

$$\frac{K}{\sigma T}=L$$

Where Lorentz number, $L=\frac{3}{2}\frac{K_B{}^2}{e^2}$.

2.1.2 Energy Band Concept of Conductor-Insulator and Semiconductor

Consider an isolated silicon atom; its energy levels are quantized. When two identical atoms are brought closer together, a quantized energy levels hybridize and split into two different levels because of the mutual interaction of the two atoms. More generally, when the N atoms are moved closer, until they reach the equilibrium inter-atomic distance d, the energy levels are split into the N levels. These N levels are very close to each other if N is large so that they eventually form a continuous energy band.

Let us now consider the silicon atoms arranged in a periodic lattice, but with a very large lattice parameter, in order to first consider each atom as isolated.

The two levels with the highest energy are labeled E1 and E2. Now let us shrink homo thetically the atom lattice: energy levels split and form two continuous bands referred as the conduction band CB and the valence band VB, Figure shows the formation of these bands as a function of the inter-atomic distance.

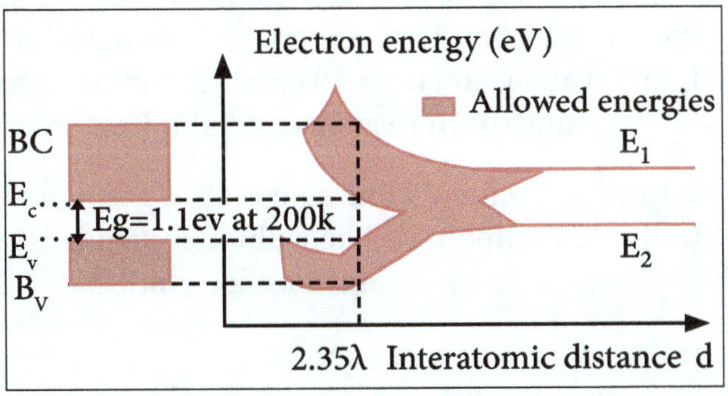

Formation of energy bands for electrons in a silicon crystal with a diamond-type lattice structure.

In a silicon crystal $\left(d_o = 2,35\,\mathring{A} \right)$, two continuous energy bands exist (CB and VB), separated by a forbidden band, which is not accessible for electrons. This forbidden region is called the gap and its width E_g is a characteristic of the material. The lowest energy level of the conduction band is denoted EC and the highest energy level of the valence band is called E_V so that we have the relationship $E_g = E_C - E_V$. The conduction and valence bands CB and VB represent the energies accessible to electrons or the energies of the states potentially occupied by electrons.

Insulator and Semiconductor

By considering their electrical properties, we can divide the materials into three groups such as,

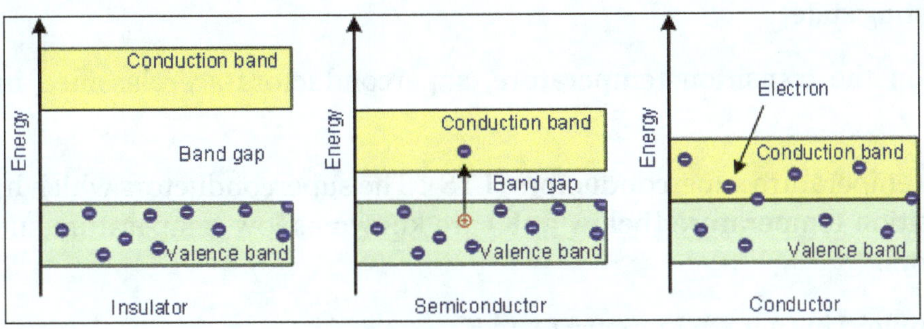

Classification of materials.

Conductors	Materials with lots of free electrons. These electrons can easily be made to flow through the material. Examples: All metals and semi-metals like carbon-graphite, antimony and arsenic.
Insulators	Materials that have very few free electrons. Examples: Plastic, glass and wood.
Semiconductors	These materials lie between the extremes of good conductors and good insulators. They are crystalline materials which are insulators when pure, but will conduct when the impurity is added and/or in response to light, heat, voltage, etc. Examples: Elements like silicon (Si), germanium (Ge), selenium (Se) and Compounds like gallium arsenide (GaAs) and indium antimonide (InSb).

2.2 Superconductor Materials: Principles of Superconductivity and Zero Resistivity

The ability of certain metals and alloys exhibit almost zero electrical resistivity when they are cooled to low temperature is known as superconducting. (i.e., maximum conductivity with zero resistance at zero Kelvin).

Each of these parameters is very dependent on the other two properties present.

Critical Temperature (T_c) or Transition Temperature

The temperature at which a normal conductor loses its resistivity and becomes a superconductor is known as critical temperature or Transition temperature. Every superconductor has its own critical temperature at which it passes over into superconducting state.

Depending on the transition temperature, superconductors are classified into two groups as:

- Low temperature superconductors (LTS): The superconductors which have low transition temperature (below 30K) are known as low temperature superconductors.

  Example: Tin (3.2 K), Mercury (4.15 K).

- High temperature superconductors (HTS): The superconductors which have high transition temperature (above 30K) is known as high temperature superconductors.

  Example: Barium - Lanthanum - Copper - Oxide (BLCO) - 35 K.

  Yttrium - Barium - Copper - Oxide - ($Y Ba_2Cu_3O_4$)-92K.

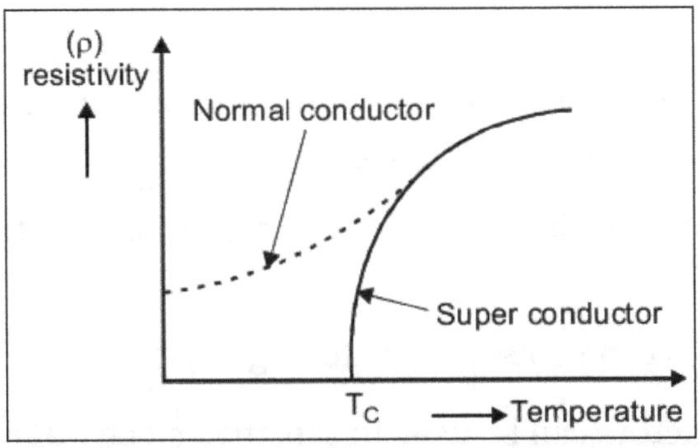

Variation of electrical resistivity with temperature.

Properties of Superconductors

At critical temperature, the following properties are observed:

- The electrical resistivity drops to zero.
- The magnetic flux lines are excluded (ejected out) from the superconductors.
- There is discontinuous change in the specific heat.
- There are small changes in the thermal conductivity and volume of the materials.

Superconductors with high values of critical temperature are called high temperature superconductors (HTSC). The discovery of copper oxide based ceramic materials by Bednorz and Muller in 1986 having critical temperature greater than 30K made a new era in the field of superconductivity.

The superconductors have the critical temperature greater than 100K are known as high superconductors or ceramic or oxide superconductors.

The superconductors due to hole states is explained by RVB theory proposed by Anderson. It is very useful for commercial and engineering applications and it is called as P-type superconductors.

Characteristics

- High T_C Superconductors have high temperatures.
- They have a modified perovskite crystal structure.

- Superconducting state is direction dependent.
- These are oxides of copper with other elements.
- These are reactive, brittle and cannot be easily modified or joined.
- For high T_C superconductors, liquid nitrogen is used instead of liquid helium.

Applications

- Electric generators can be made by using superconductors with smaller size, less weight and low energy consumption.
- Superconductors can be used for the transmission of power over very long distances.
- Superconductors can be used in switching devices.
- The superconductors can be used in sensitive electrical instruments.
- It can be used as a memory or storage element in computers.
- These are used to design Cryotron, Maglev, Josephson Devices and SQUID.
- DC superconducting motors are used in ship propulsion and in large mills.
- Superconducting magnetic field may be used to launch satellite into orbit directly from the earth without use of rockets.
- Separation can be done by using machines made of superconducting magnets.
- These are used in NMR (Nuclear Magnetic Resonance) imaging equipments which is used for scanning purposes.
- Superconductors are used for the detection of brain tumour, defective cells, etc.
- Superconducting solenoids are used in magneto hydrodynamic power generation to maintain the plasma in the body.

SQUIDs (Superconducting Quantum Interference Devices)

SQUID stands for Superconductors Quantum Interference Device. It is a double junction quantum interferometer formed from two Josephson junctions mounted on a superconducting ring. SQUID is based on the flux quantization in a superconducting ring.

The total magnetic flux passing through the ring is quantized. It is used to detect very minute magnetic field of the order of 10^{-14} tesla.

SQUID consists of a superconducting ring which can have the magnetic field of quantum values (1,2,3,....) of flux placed in between two Josephson junctions as shown in the figure.

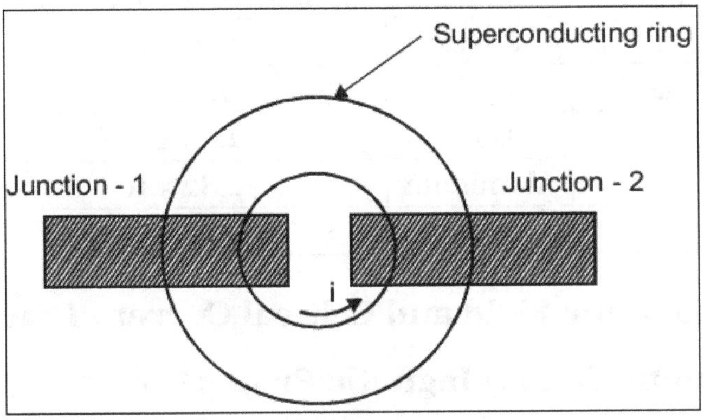

SQUID.

When the magnetic field is applied perpendicular to the plane of the ring, the current is induced at the two Josephson junctions.

The induced current produces the interference pattern and if flows around the ring so that the magnetic flux in the ring can have the quantum value of magnetic field applied.

Applications

- SQUID can be used to detect the variation of very minute magnetic signals in terms of quantum flux.
- It is used as a storage device for magnetic flux.
- It is used to study earthquakes and to remove paramagnetic impurities.

Zero Resistivity

It the property of material which resists the flow of electric current through material. It is the reciprocal of conductivity.

It is dented by 'ρ'. Resistivity of a material of a conductor can be determined as below,

$$\rho = R\frac{A}{l}$$

Where,

'R' is the resistance of conductor in Ω.

'A' is the cross sectional area of conductor in m^2.

'l' is the length of the conductor in meter.

SI unit of resistivity of is Ω-meter. Resistivity of some materials is listed below:

SL.NO	ELEMENT	Resistivity at 20°C in Ω - m
1	Silver	1.59×10^{-8}
2	copper	1.7×10^{-8}
3	Gold	2.44×10^{-8}
4	Aluminum	2.82×10^{-8}
5	Iron	1.0×10^{-8}

2.2.1 Critical Magnetic Field and Critical Current Density

1. Critical Magnetic Field (Magnetic Property)

A very strong magnetic field is applied to superconducting material and it disappears superconducting property is called as critical magnetic field,

$$H_C = H_o \left[1 - \left[\frac{T}{T_C} \right]^2 \right]$$

Where,

H_o = Critical field at 0K.

It is noted that when the temperature of a material increases, the value of critical magnetic field decreases. Therefore, the value of critical magnetic fields are different for different materials.

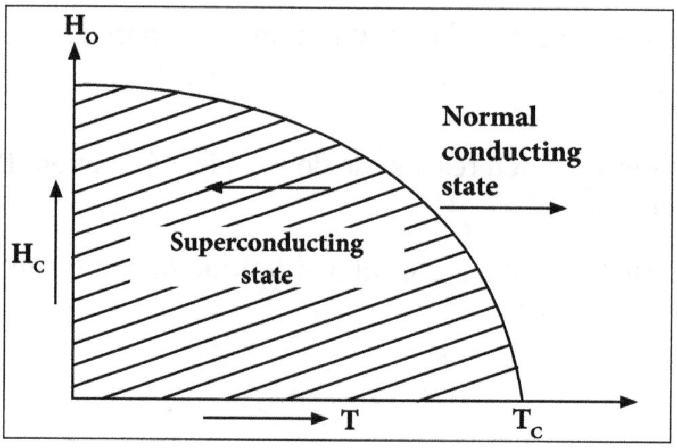

Critical magnetic field.

2. Diamagnetic Property (Meissner Effect)

If a normal conducting material is placed in a magnetic field of flux density B, the magnetic lines penetrate through the material.

Now the material is cooled below its transition temperature when T ≤ TC, then the magnetic lines of forces are expelled out from the material as shown in the figure.

We know that, a diamagnetic material have the tendency to expel the magnetic lines of force. Since the superconductor also.expels the magnetic lines of force and it behaves as a perfect diamagnet. This behaviour was first observed by Meissner and hence called as Meissner effect.

I. Meissner Effect

When the superconducting material is subjected to a uniform magnetic field, under the condition $T \leq T_C$ and $H \leq H_C$, the magnetic flux lines are excluded from the material. Thus, the material exhibits perfect diamagnetism. This phenomenon is called as Meissner effect.

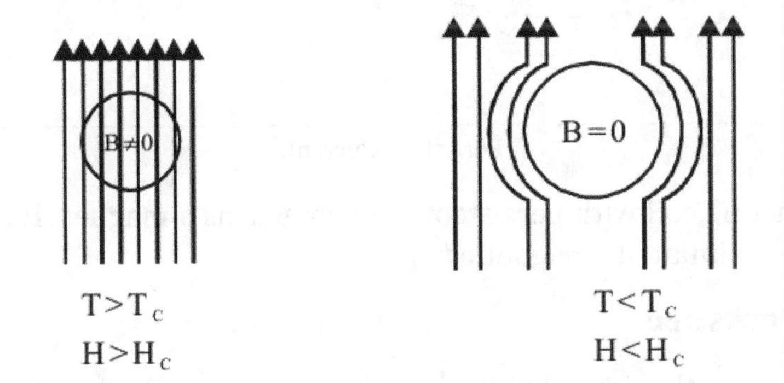

(a) Normal state **(b) Super conducting state**

Meissner effect.

Application of Meissner Effect

- It is a standard test to prove whether the material is a perfect superconductor or not.
- This effect is used for magnetic levitated train.

3. Effect of Heavy Current

The superconducting property disappears when a heavy current flows, since current flow will set up a magnetic field.

According to Silsbee's rule, for a superconducting wire, the induced current to destroy the superconducting property is given by, $I_C = 2\, r\, H_C$.

Where,

r is the radius of the superconductor.

H_C is the critical magnetic field.

4. Persistence of Current

Persistence current is one of the most important properties of a superconductor. When a current of large magnitude is induced in a superconducting ring, the current persist in the ring even after the removal of the field at the temperature below the critical temperature, such a current flow without reducing its strength is known is persistent current.

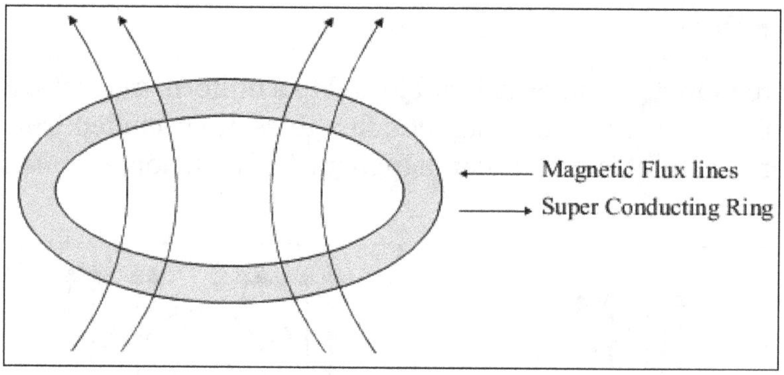

Persistent Current.

The superconducting coil with persistent current acts as a magnet. It does not require power supply to maintain its magnetic field.

5. Effect of Pressure

If pressure increases, the critical temperature also increases. Therefore, certain materials are brought into the superconducting state by increasing the pressure. Research is going on to get the superconducting state at room temperature by applying heavy pressure.

6. Isotope Effect

The presence of isotopes in superconductor change the transition temperature of the superconductor. The transition temperature is found to be inversely proportional to the square root of the atomic weight of the isotope (M).

$$T_C \propto \frac{1}{\sqrt{M}}$$

$$T_C = \frac{\text{constant}}{\sqrt{M}}$$

$$\sqrt{M} \times T_C = \text{Constant}$$

The transition temperature of the heavier isotope is less than that of the lighter isotope.

General properties

- There is no change in elastic properties, photo electric properties and crystal structure.
- The transition temperature is unchanged with the frequency variation.

2.2.2 Type I and II Superconductors

Based on the value of HC we have:

- Type I or Soft superconductors.
- Type II or Hard superconductors.

Type I Superconductor

In type I superconductor, the magnetic field is completely excluded from the material below the critical magnetic field and the material loses its superconducting property abruptly.

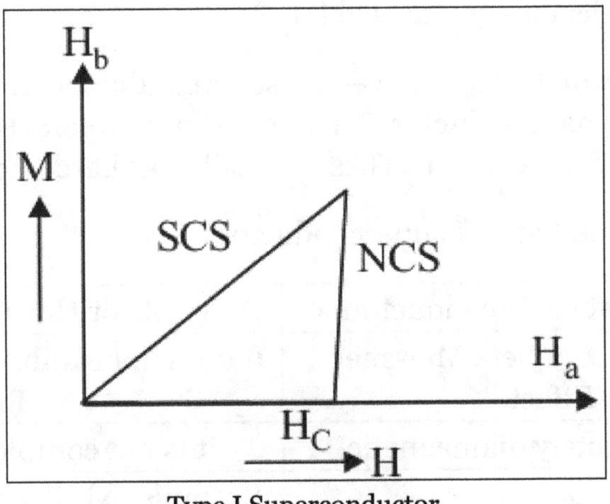

Type I Superconductor.

Characteristics

- They exhibit complete Meissner Effect.
- They have only one critical magnetic field value.
- Below, the material behaves as superconductor and above, the material behaves as normal conductor. These are called as Soft superconductors.

Type II Superconductor

In type II superconductor, the magnetic field is excluded from the material and the material loses its superconducting property gradually rather than abruptly.

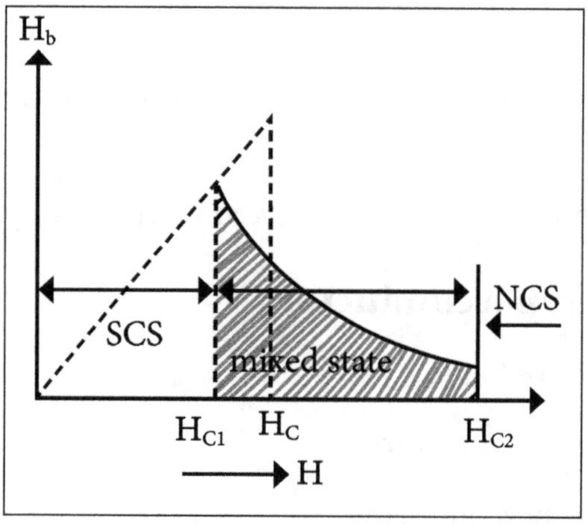

Type II Superconductor.

Characteristics

- They do not exhibit a complete Meissner Effect.
- They have two critical magnetic field values. Lower critical magnetic field [H_{C1}] and Higher critical magnetic field [H_{C2}].
- Below H_{C1}, the material behaves as superconductor and above, the material behaves as normal conductor. The region in between [H_{C1}] and [H_{C2}] is called mixed state or vortex region. These are called as hard superconductors.

Difference between Type I and II superconductors:

S .No.	Type I or Soft superconductors	Type II or Hard superconductors
1.	It exhibits complete Meissner Effect.	It does not exhibit a complete Meissner Effect.
2.	It is completely diamagnetic.	It is not completely diamagnetic.
3.	Example: Tin, Lead, Mercury, etc.,	Example: Nb-Zr, Nb-Ti, Nb-Sn, Va-Ga, etc.,
4.	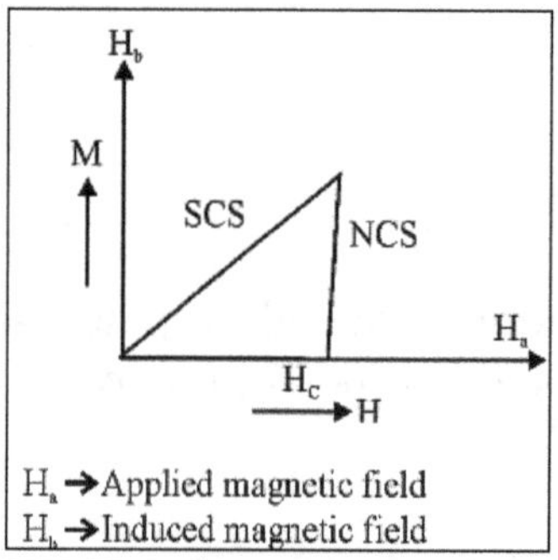	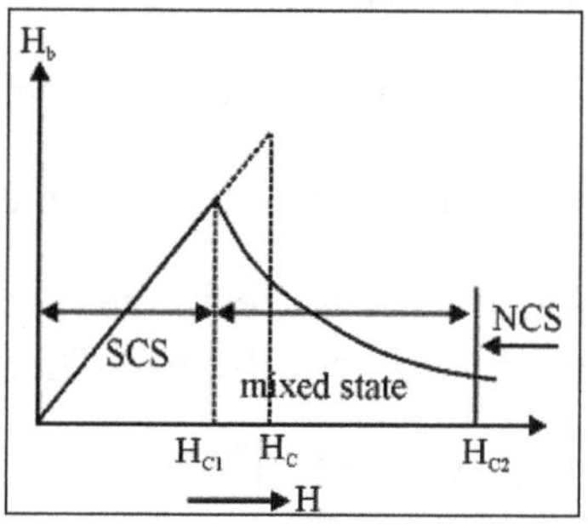
5.	The value of H_C is always too low. It is about 0.1 tesla.	The values of H_C is high and it is about 30 tesla.

6.	It has only one critical magnetic field value.	It has two critical magnetic field values. Lower critical magnetic field [H_{C1}] and Higher critical magnetic field [H_{C2}].
7.	No mixed state exists.	Mixed state is present.
8.	These are called Soft superconductors.	These are called Hard superconductors.
9.	It requires low magnetic field to destroy the superconductivity.	It requires large magnetic field to destroy the superconductivity.
10.	It loses its magnetization abruptly.	It loses its magnetization gradually.

2.2.3 Applications of Superconductors

Superconductors are already used in many fields: electricity, medical applications, electronics and even trains. They are used in laboratories, especially in particle accelerators, in astrophysics with the use of bolometers, in ultrasensitive magnetic detectors called SQUIDs and in superconducting coils to produce very strong magnetic fields.

However, they need to be cooled to very low temperatures, which restricts their use in our everyday life. But new applications are already operational in laboratories and will be able to spread to our cities and our homes if the cooling process becomes less expensive or, better, if we discover superconducting materials that do not require any cooling. If this happens, we can expect an actual revolution in energies and environment on the one side and transportation and computer science on the other.

Electric Power

Superconductors enable a variety of applications to aid our aging and heavily burdened electric power infrastructure - for example, in generators, transformers, underground cables, synchronous condensers and fault current limiters. The high power density and electrical efficiency of the superconductor wire results in highly compact, powerful devices and systems that are more reliable, efficient and environmentally benign.

Transportation

A rapid and efficient movement of people and goods, by the land and by sea, poses very important logistical, environmental, land use and the other challenges. Superconductors are enabling a new generation of transport technologies including ship propulsion systems, magnetically levitated trains and railway traction transformers.

Medicine

This advances in HTS promise more compact and less costly Magnetic Resonance Imaging (MRI) systems with superior imaging capabilities. In addition, the Magnetic Source Imaging (MSI), Magneto-Encephalography (MEG) and Magneto Cardiology (MCG) enable a non-invasive diagnosis of brain and heart functionality.

Industry

Large motors rated at 1000 HP and above consume 25% of all the electricity generated in the United States. They offer a prime target for the use of HTS in substantially reducing the electrical losses. Powerful magnets for materials purification, water remediation and industrial processing are also in the demonstration stages.

Communications

Over the past decade, the HTS filters have come into widespread use in cellular communications systems. They enhance the signal-to-noise ratios, enabling reliable service with fewer, more widely-spaced cell towers. As the world moves from analog to all digital communications, LTS chips offer dramatic performance improvements in many commercial and military applications.

Scientific Research

Using the superconducting materials, today's leading-edge scientific research facilities are pushing the frontiers of human knowledge and pursuing breakthroughs that could lead to new techniques ranging from the clean, abundant energy from nuclear fusion to computing at speeds much faster than the theoretical limit of silicon technology.

2.3 Dielectric Materials

When a neutral atom is placed in an external de electric field E, the positively charged, massive nucleus is pushed in the direction of the electric field while the negatively charged but much lighter electrons are pushed in the opposite direction through a much longer distance.

Hence the centres of positive and negative charges in the atom no longer coincide and they undergo net relative dis-placement, r. The atom which was originally electrically neutral now acquires an electric dipole moment p = (Ze) r. It can be assumed that, to first order, the induced dipole moment is proportional to E and so the dipole moment can be written as,

$$p = \alpha E$$

A is known as the polarizability of the atom (or molecule).

Dielectric

Dielectric are the materials which are transferable and by applying the certain conditions. They can be transformed into the polar conductor. But in the real time situation, they are insulator which does not allow the flow of electrons from them due to their atomic structure and strong cohesive forces among the items which resist any change in their structure. Dielectric materials can be mold to conduct the current flow through them.

Dielectric Properties

- Insulating material is a very poor conductor of electric current.
- Dielectrics have no loosely bound electrons and thus no current flows through them.
- When they are placed in the electric field, the positive and negative charges within the dielectric are displaced minutely in the opposite directions, which reduces the electric field within the dielectric. Examples of dielectrics includes plastics, glass and ceramics.

Dielectrics in Static Electric Field

The presence of an external electric field causes a force to be exerted on each charged particle and results in small displacements of positive and negative charges in opposite directions. These displacements polarize a dielectric material and create electric dipoles, which equivalent to an volume charge density ρp.

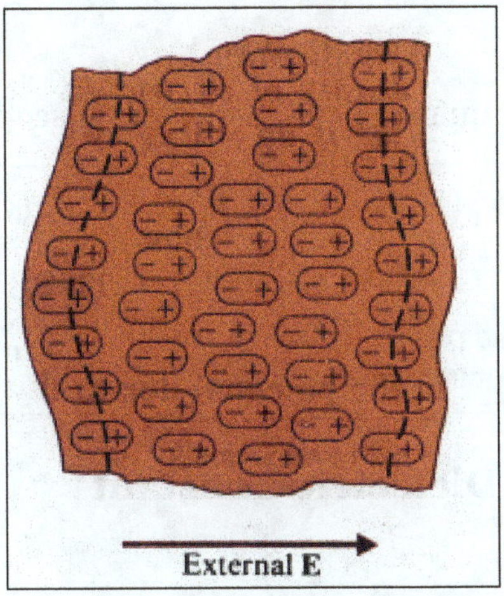

Polarized dielectric medium.

The polarized charge density is related to the polarization vector P by,

$$\rho_p = -\nabla \cdot \overline{P}$$

The "–" sign means that the field generated by the electric dipoles is in the opposite direction of external electric field E. The divergence postulation must be modified to include the effect of ρ_p,

$$\nabla \cdot \vec{E} = \frac{1}{\varepsilon_o}\left(\rho + \rho_p\right)$$

$$\rho_p = -\nabla \cdot \vec{P}$$

or,

$$\nabla \cdot \left(\varepsilon_0 \vec{E} + \vec{P}\right) = \rho$$

We now define a new fundamental field quantity, the electric flux density or electric displacement D is given as,

$$\vec{D} = \varepsilon_0 \vec{E} + \vec{P} \qquad \left(C/m^2\right)$$

or,

$$\nabla \cdot \vec{D} = \rho \qquad \left(C/m^3\right)$$

When the dielectric properties of the medium are linear and isotropic then the polarization is directly proportional to the electric field intensity and the proportionality constant is independent of the direction of the field.

That is,

$$\vec{P} = \varepsilon_0 \chi_e \vec{E}$$

Where X_e is a dimensionless quantity called electric susceptibility.

$$\vec{D} = \varepsilon_0(1+\chi_e)\vec{E} = \varepsilon_0\varepsilon_r\vec{E} = \varepsilon\,\vec{E} \quad (C/m^2)$$

Relative permittivity or dielectric constant

Absolute permittivity

2.3.1 Polarization and Dielectric Constant

Polarization

A material may have several dielectric polarization effects that contribute to its overall permittivity. On application of an electric field to a dielectric material the opposite electric charges are displaced. These charges become polarized to compensate for the electric field such that the positive and negative charges move in the opposite directions. The maximum polarization Po occurs at zero frequency. Because of inertia of the charge carriers, the polarization process requires time, therefore the polarization P is in general not in-phase with the electric field E. A frequency dependent complex permittivity "k = k"– jk is used to capture both amplitude and phase information,

$$P = k \in 0\, E$$

The spectra of polarization mechanisms have the form of relaxation or resonance. If the charge is moved against a restoring force a resonance spectrum occurs.

Dielectric Constant

Dielectric constant is the property of the electrical insulating material equal to the ratio of the capacitance of the capacitor filled with the given material to the capacitance of an identical capacitor in a vacuum without the dielectric material. The insertion of the dielectric between the plates of a parallel-plate capacitor always increases its capacitance or the ability to store the opposite charges on each plate, compared with this ability when the plates are separated by the vacuum.

If C is the value of the capacitance of a capacitor filled with the given dielectric and C_o is the capacitance of the identical capacitor in a vacuum, the dielectric constant, symbolized by the Greek letter kappa, κ can be simply expressed as $\kappa = C/C_o$. Dielectric constant is a number without dimensions. It denotes the large-scale property of the dielectrics without specifying the electrical behaviour on the atomic scale.

The value of the static dielectric constant of any material is always greater than one, its value for a vacuum. The value of the dielectric constant at the room temperature (25° C or 77° F) is 1.00059 for air, 2.25 for paraffin, 78.2 for water and about 2,000 for barium titanate (BaTiO3) when the electric field is applied perpendicularly to the principal axis of the crystal. Because the value of the dielectric constant for air is nearly the same as that of the vacuum, air does not increase the capacitance of the capacitor.

Dielectric constants of liquids and solids is determined by comparing the value of the capacitance when the dielectric is in place to its value when the capacitor is filled with air. The dielectric constant is sometimes referred as relative permittivity or specific inductive capacity.

A parallel plate capacitor with plate area 'A' and separation between plates 'd' capacitance of an air capacitor $Co = Q/V = A\,\varepsilon o/d$. If a dielectric is inserted then $C = Q/V = \varepsilon o\ \varepsilon r\ A/d$.

Hence,

$$C = \varepsilon r x C_o$$

- It indicates that insertion of a dielectric increases the capacitance of a capacitor and measuring A, d & C we can measure εr.
- If an air capacitor and a dielectric capacitor are given the same charge 'Q', then potential on the 2nd is, $V = V_o/\varepsilon_r$.
- 'ε_r' is called the dielectric constant of the dielectric.
- $\varepsilon_r = \varepsilon/\varepsilon_o$, is also called as relative permittivity.
- $\varepsilon_o = 8.85 X\ 10^{-12}$ F/m.

Dielectric Strength

Dielectric Strength of a substance is the total potential difference between the two points of a substance per unit thickness of the medium; up to which the substance can withstand the electric field or up to which no Electric breakdown occurs in the substance. The dielectric strength is given by the potential gradient necessary to cause breakdown of an insulator or substance.

The standard unit of Dielectric Strength is Volt per Meter (V/M), which can be mostly used as kilo Volt per millimeter (kV/mm).

Dielectric Strength of Air

The dielectric strength of Air is 3kV/mm, Which means that is two electrodes are placed one millimeter away from each other in air, then the air acts as a insulator until the voltage between the electrodes reaches the level of 3 kilo volts or 3000 volts, after which the air starts behaving as a good conductor which causes sparks and fire.

2.4 Dielectric Susceptibility

Electric susceptibility (χ_e)

Electric susceptibility is defined as the ratio of polarization P to electric field strength E. The polarization vector (P) is proportional to the total applied electric field intensity (E) and is in the same direction of E. Therefore, the polarization vector can be written as,

$$P = \varepsilon_o \chi_e E$$

Where χ_e is electric susceptibility.

Dielectric Constant or Relative Permittivity (ε_r)

The dielectric characteristics of a material are determined by the dielectric constant or relative permittivity of the material.

It is defined as the ratio of the permittivity of the medium to the permittivity of the free space,

$$\varepsilon_r = \frac{\text{Permitivity of the medium}}{\text{Permitivity of free space}}$$

$$\varepsilon_r = \frac{\varepsilon}{\varepsilon_o}$$

Where,

$$\varepsilon_o = 8.854 \times 10^{-12} \text{ F/m}.$$

Relation between electric susceptibility (χ_e) and relative permittivity (εr),

We know that,

$$P \propto E$$

$$P = \varepsilon_o \chi_e E$$

$$\chi_e = \frac{P}{\varepsilon_o E}$$

But,

$$\varepsilon_r = 1 + \chi_e$$

$$\chi_e = \varepsilon_r - 1$$

$\therefore$ We can write $P = \varepsilon_o E(\varepsilon_r - 1)$

It is a measure of characteristic of every dielectric.

2.4.1 Polarization Mechanisms

Polarization Mechanisms Involved in a Dielectric Material

There are four different types of polarization:

- Electronic or induced polarization.
- Ionic or atomic polarization.
- Orientation or dipolar polarization.
- Space - Charge or interfacial polarization.

Electronic Polarization

Electronic Polarization occurs due to the displacement of the positively charged nucleus and negatively charged electron in the opposite directions by an external electric field. This creates a dipole moment in the dielectric. This is known as electronic polarization.

The induced dipole moment,

$$\mu_e = \alpha_e E \qquad \dots (1)$$

Where,

α_e - Electronic polarisability.

E - Electric field Strength.

It is proportional to volume of the atoms and is independent of temperature.

Calculation of Electronic Polarization A_e

Consider an atom of a dielectric material of nuclear charge (Z_e). The electrons of charge ($-Z_e$) are distributed uniformly throughout the atom (sphere) of radius R as shown in figure.

Without Field

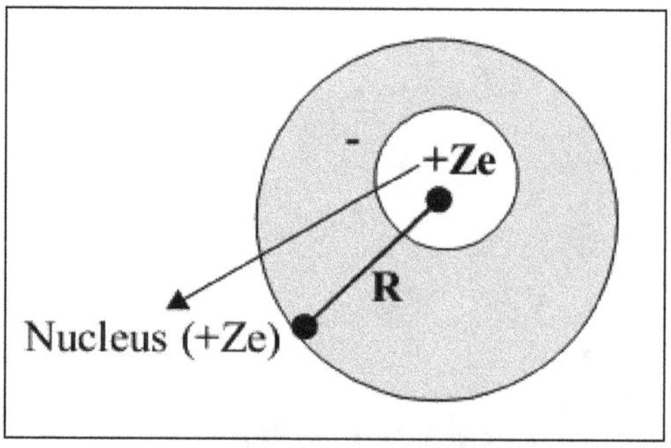

Electronic polarization without field.

$$\left.\begin{array}{l}\text{Charge of the charged}\\ \text{sphere (electrons) of radius R}\end{array}\right\} = \frac{\text{Total charge of sphere with radius R}}{\text{Volume}}$$

$$= \frac{-Ze}{\frac{4}{3}\pi R^3}$$

$$= \frac{-3Ze}{4\pi R^3} \qquad \text{...(1)}$$

With Field

When the atom of the dielectric is placed in an electric field (E), two types of forces arise:

- Lorentz force: Force which separates electrons and positive nucleus due to applied field.
- Coulomb force: An attractive force which is produced after separation.

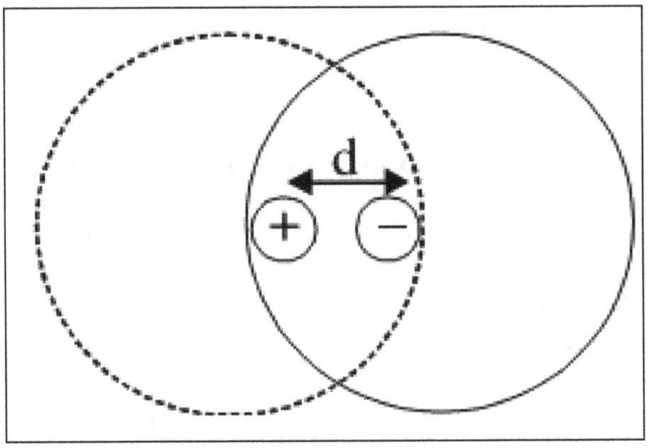

Electronic polarization with field.

At Equilibrium

Lorentz force = Coulomb force.

Lorentz force = (charge of the electron) × (intensity of applied Field),

$$=(-Z_e)\ (E) \quad \text{... (2)}$$

$$\text{Coulomb force} = \frac{\begin{bmatrix}\text{Charge of the}\\ \text{positive nucleus}\end{bmatrix} \times \begin{bmatrix}\text{Total negative charges}\\ \text{enclosed in the sphere}\end{bmatrix}}{4\pi\varepsilon_0 d^2} \quad \text{...(3)}$$

$$\left.\begin{matrix}\text{The total number of}\\ \text{negative charges enclosed}\\ \text{in the sphere of radius 'd'}\end{matrix}\right\} = \begin{bmatrix}\text{Charge denstity}\\ \text{of electrons}\end{bmatrix} \times [\text{volume of the sphere}]$$

$$= \left[\frac{-3Ze}{4\pi R^3}\right]\left[\frac{4\pi d^3}{3}\right]$$

$$= \frac{-Zed^3}{R^3} \quad \text{...(4)}$$

Substitute equation (4) in equation (3),

$$\text{Coulomb force} = \frac{Ze \times \left[\frac{-Zed^3}{R^3}\right]}{4\pi\varepsilon_0 d^2}$$

$$= \frac{-Z^2 e^2 d}{4\pi\varepsilon_0 R^3}$$

$$\text{Coulomb force} = \frac{-Z^2 e^2 d}{4\pi\varepsilon_o R^3} \qquad ...(5)$$

At equilibrium, Coulomb force and Lorentz force must be equal and opposite.

Lorentz force (equation (2)) = Coulomb force (equation (5)),

$$(-Ze)(E) = \frac{-Z^2 e^2 d}{4\pi\varepsilon_o R^3}$$

$$E = \frac{Zed}{4\pi\varepsilon_o R^3}$$

$$d = \frac{4\pi\varepsilon_o R^3 E}{Ze} \qquad ...(6)$$

The displacement (d) of electron cloud is proportional to applied field (E).

Now the two electric charges +Ze and –Ze are displaced by a distance d under the influence of the field and form an induced dipole moment which is given by,

Induced dipole moment (μe) = Charge of the electron × displacement,

$$\mu_e = Z_e d$$

Substituting the value of 'd' from equation (5),

$$\mu_e = \frac{Ze\,\pi\varepsilon_o R^3 E}{Ze}$$

$$\mu_e = 4\pi\varepsilon_o R^3 E$$

$$\mu_e \,\alpha\, E$$

$$\mu_e = \alpha_e E$$

Where $\alpha_e = 4\Pi\varepsilon_o R^3$ is called electronic polarization which is proportional to volume of the atom.

Conclusion:

- Electronic polarization is independent of temperature.
- Electronic polarization occurs in all dielectric materials.
- It is proportional to the volume of atoms in the material.

Electronic Polarization in Terms of E_o And E_r

P_e is the total number of dipoles per unit volume [V = 1],

$$P_e = \frac{N\mu_e}{V} = N\mu_e$$

Where N is the number of atoms per m³,

But,

$$\mu_e = \alpha_e E$$

$$P_e = N\alpha_e E$$

We know that,

$$P_e = \varepsilon_o E(\varepsilon_r - 1)$$

$$N\alpha_e E = \varepsilon_o E(\varepsilon_r - 1)$$

$$\alpha_e = \frac{\varepsilon_o(\varepsilon_r - 1)}{N}$$

This is called electronic Polarizability.

Ionic Polarization

Polarization which arises due to the displacement of cations (+ve) and anions (-ve) in opposite directions and occurs in ionic solids in the presence of electric field, is called ionic polarization.

Example: NaCl, KCl crystal.

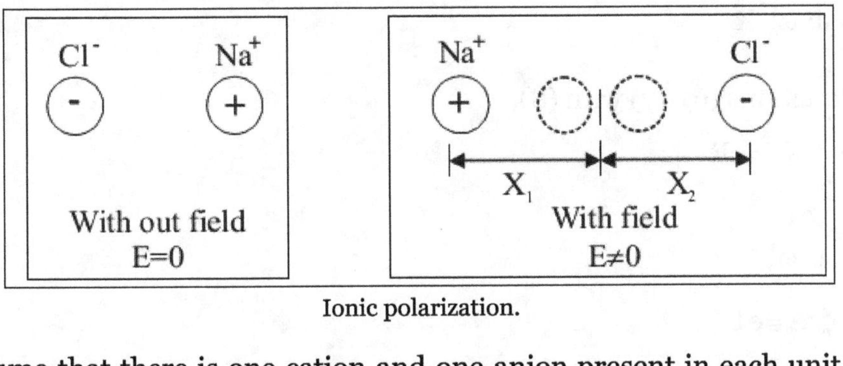

Ionic polarization.

Let us assume that there is one cation and one anion present in each unit cell of the ionic crystal (Na cl). When an electric field is applied, let X_1 and m X_2 be the distance to which positive and negative ions move from their equilibrium positions.

The resultant dipole moment per unit volume due to ionic displacement is given by,

Dipole moment μi = Charge × Displacement

$$\mu_i = e\left[x_1 + x_2\right] \qquad ...(1)$$

When the field is applied, the restoring force produced is proportional to the displacement.

For +ve ion, Restoring force $F \infty X_1$

$$F = \beta_1 X_1 \qquad ...(2)$$

For –ve ion, restoring force $F \infty X_2$

$$F = \beta_2 X_2 \qquad ...(3)$$

Here β_1 and β_2 are restoring force constants, which depends on the mass of the ions and angular frequency of the molecule in which ions and present.

If m is mass of positive, ion, M is mass of negative ion and ωo is angular frequency we can write,

$$\beta_1 = m\,\omega_o^2 \qquad ... (4)$$

$$\beta_2 = M\,\omega_o^2 \qquad ...(5)$$

Equation (2) and (3) can be written as,

$$F = m\,\omega_o^2\,X_1 \qquad ...(6)$$

$$F = M\,\omega_o^2\,X_2 \qquad ...(7)$$

We know,

$$F = m\,\omega_o^2\,X_1 \qquad ... (8)$$

Equation equations (6), (7) with (8),

$$m\,\omega_o^2\,X_1 = eE$$

$$x_1 = \frac{eE}{m\omega_o^2} \qquad ...(9)$$

$$M\omega_o^2 x_2 = eE$$

$$x_2 = \frac{eE}{M\omega_o^2} \qquad ...(10)$$

$$x_1 + x_2 = \frac{eE}{\omega_0^2}\left[\frac{1}{m} + \frac{1}{M}\right] \quad ...(11)$$

Substitute equation (11) in equation (1),

$$\mu_i = \frac{e^2 E}{\omega_0^2}\left[\frac{1}{m} + \frac{1}{M}\right] \quad ...(12)$$

We know that,

$$\mu_i = \alpha_i E$$

Where αi is ionic polarizability,

$$\alpha_i = \frac{e^2}{\omega_0^2}\left[\frac{1}{m} + \frac{1}{M}\right]$$

Conclusion:

- Valent polarizability (αi) is inversely proportional to the square of angular frequency of the ionic molecule.
- It is independent of temperature.
- It is directly proportional to its reduced mass given by, $\left[\frac{1}{m} + \frac{1}{M}\right]$.

Orientation Polarization

The orientation polarization is due to the existence of a permanent dipole moment (polar molecule) in the dielectric medium. Polar molecules have permanent dipole even in the absence of an electric field.

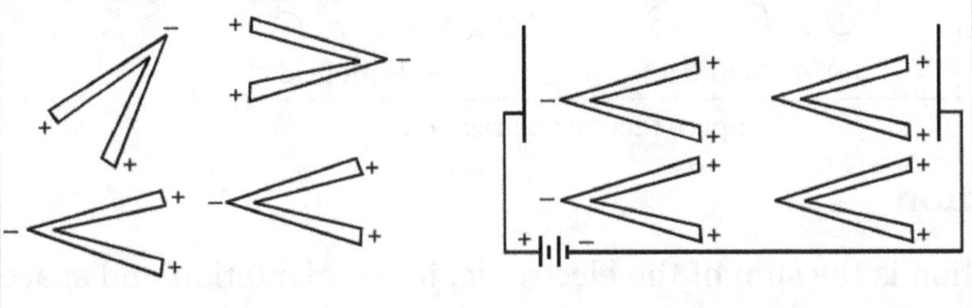

(a) Without field (b) With field

Orientation polarization.

When an electric field is applied on the dielectric medium with polar molecules electric field tries to align these dipoles along its field direction as shown in figure. Due to here is a resultant dipole moment in that material and this process is called orientation polarization.

This polarization depends on temperature. When the temperature is increased, the dipoles are aligned in random directions,

Orientation polarization, $P_O = \frac{N\mu^2 E}{3KT}$

Where,

N - Number of atoms.

Orientation polarization, $P_o = N\alpha_o E$

$$\alpha_o = \frac{P_o}{NE}$$

i.e., αo is inversely proportional to temperature of the material.

Space Charge Polarization

The space charge polarization occurs due to the diffusion of ions along the field direction giving rise to redistribution of charges in the dielectric.

Normally this type of polarization occurs in ferrites and semiconductors and it is very small when compared to other types of polarization.

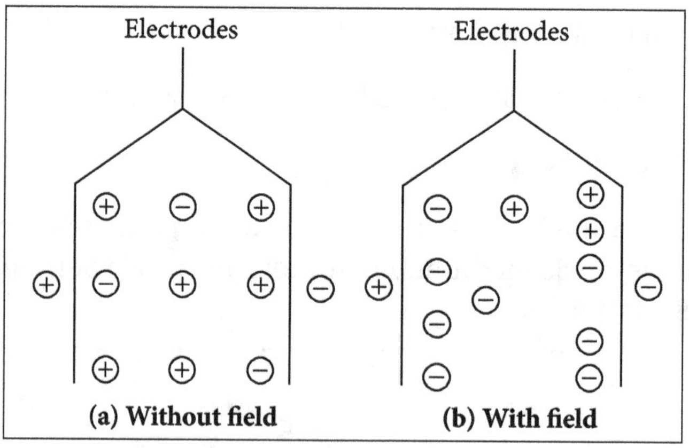

Space Charge polarization.

Total Polarization

The total polarization is the sum of the electronic, ionic orientation and space charge polarization.

Orientation polarization, $P_o = N\alpha_o E$

Total polarization $P_{total} = P_e + P_i + P_o + P_s$

The total polarizability $\alpha = \alpha_e + \alpha_i + \alpha_o + \alpha_s$

Here αs is very small and it can be neglected,

$$\alpha = \alpha_e + \alpha_i + \alpha_o$$

$$\alpha = 4\pi\varepsilon_0 R^3 + \frac{e^2}{\omega_0^2}\left[\frac{1}{m} + \frac{1}{M}\right] + \frac{\mu^3}{3KT}$$

$$P = NE\alpha$$

$$P = NE\left[4\pi\varepsilon \ R \ + \frac{e^2}{}\left[\frac{1}{m} + \frac{1}{M}\right] + \frac{3}{3KT}\right]$$

This equation is known as Langevin - Debye equation.

2.4.2 Temperature and Frequency Dependence of Dielectric Constant

Frequency Dependence

On application of an alternating field across the material, the polarization occurs as function of time,

üüüüü(üü)i= $_m[- \quad [- \quad _r]]$

Where P is the maximum polarization attained due to applied field and t is the relaxation time which is the time taken for a polarization process to reach 0.63 of the maximum value. The relaxation times are different for different kinds of polarization mechanisms.

- Electronic polarization is very fast and is completed at any instant of time even when the frequency of the voltage is very high in the optical range (1015Hz). Thus, it occurs at all frequencies.
- Ionic Polarization is slower and the ions do not respond when the voltage corresponds to visible optical frequencies, i.e., the electric field changes in polarity at very fast, so that the ions are not able to reorient themselves up to the field. So the ionic polarization does not occur at visible optical frequencies. It occurs only at frequencies less than 1013 Hz.
- Orientation Polarization is even slower than ionic polarization and occurs only at electrical frequencies (audio and radio frequencies 106Hz).
- Space-charge polarization is the slowest process because the ions have to diffuse (jump) over several interatomic distances. This occurs at very low frequencies of 50 - 60 Hz (power frequencies).

Thus at low frequencies, all the four polarizations will occur and the total polarization is very high but at high frequencies, the value of the total polarization is very small.

The following graphs show the frequency dependence of polarization mechanism and the corresponding power losses at those frequencies.

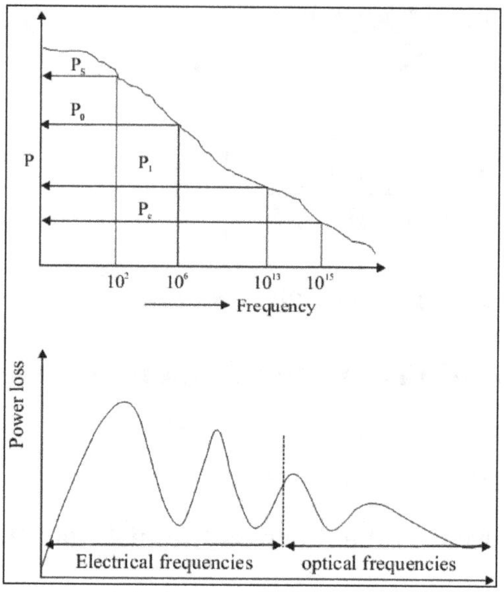

Polarization Mechanisms.

Temperature Dependence

Electronic and ionic polarizations are independent of temperature and the orientation and space charge polarizations are dependent on temperature. Orientation polarization is inversely proportional to the temperature.

Orientation polarization decreases when temperature increases. Because the random nature decreases, the permanent dipoles align along the field direction. Hence, the dielectric constant increases.

Space charge polarization is directly proportional to the temperature. The space charge polarization increases with increase of temperature. It is because of the fact that the thermal energy helps to overcome the activation barrier and the ions diffuse easily, this results in decrease of dielectric constant.

2.4.3 Dielectric Breakdown

When a dielectric material loses its property and permits the flow of a large current, it is said to be dielectric breakdown.

Types of dielectric breakdown:

- Intrinsic breakdown.

- Thermal breakdown.
- Electrochemical breakdown.
- Defect breakdown.
- Discharge breakdown.

1. Intrinsic Breakdown

When a dielectric material is subjected to large electric field, large number of electrons are transferred from the valence band to the conduction band. Thus, the dielectric material loses its insulating property and becomes a conductor. This is known as intrinsic breakdown.

Further the conducting electrons may collide with the atoms and release some of the valence electrons. These electrons may collide with some more atoms and may release more electrons. This becomes a chain process resulting in a large current. It is known as avalanche breakdown.

Characteristics:

- This type of breakdown occurs at room temperature and low temperatures.
- This requires relatively large electric fields.
- This kind of breakdown occurs in the materials.

2. Thermal Breakdown

Thermal breakdown occurs in dielectric when the rate of heat generation is greater than the rate of heat dissipation.

When a dielectric is subjected to an electric field, heat is generated. The generated heat is dissipated by the dielectric material. When the amount of heat generation is higher than the amount of heat dissipation, the temperature inside the dielectric increases and this causes the breakdown called thermal breakdown.

Characteristics:

- This can occur only at high temperatures.
- The thermal breakdown time is of the order of few milliseconds.
- This requires moderate electric fields.
- It depends on size and shape of dielectric material.
- Since the dielectric loss is directly proportional to frequency, the electric field strength to create this dielectric breakdown will be smaller for AC fields and higher DC fields.

3. Discharge Breakdown

This type of breakdown occurs when the insulator contains occluded gas bubbles. When the dielectric is subjected to an electric field, the gas present in the material will be easily ionized than the solids.

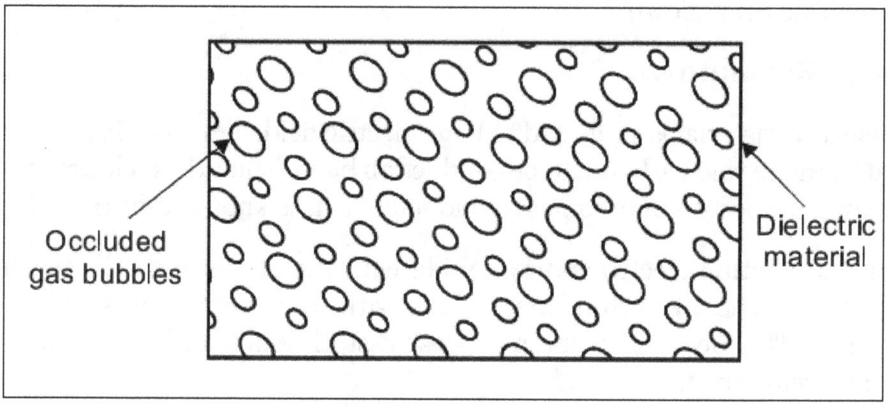

Discharge breakdown.

The ionized gas particles bombarded with the solid dielectric and produces large ionization current called discharge breakdown.

Characteristics

- This is possible at low voltages where there are large number of occluded gas bubbles is present in the material.
- The life time of the material depends upon the number of discharge taking place inside the material.
- This occurs due to presence of gas bubbles.

4. Electrochemical Breakdown

When temperature increases, mobility of ions increases and hence leakage current also increases. This decreases the insulation resistance and finally creates a dielectric breakdown. Hence this type of breakdown is called electrochemical breakdown.

Characteristics:

This is used to determine the leakage current, density of ions and dipoles inside the material.

- They are accelerated by high temperatures.
- This occurs only at low temperature.
- This type of breakdown occurs even in the absence of electric field also.

Defect Breakdown

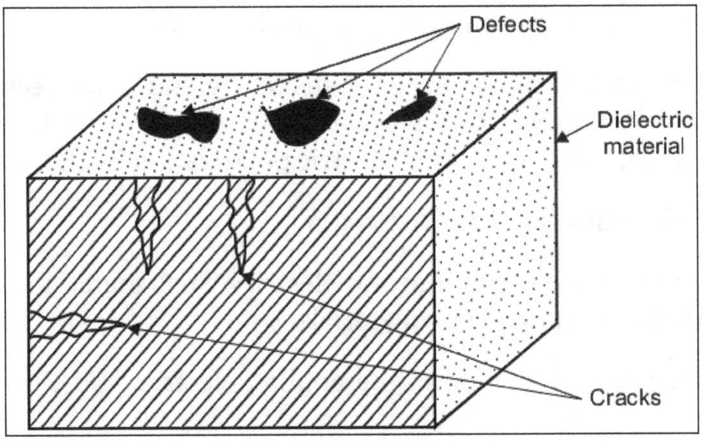

Defect breakdown.

If the surface of the dielectric material has defects such as cracks, pores etc., moisture and other impurities can fill at these places leading to breakdown, this type of breakdown is called defect breakdown.

Remedies to Avoid Breakdown Mechanisms

To avoid breakdown, the dielectric material should have the following properties:

- It should have high resistivity.
- It must have low dielectric loss and less density.
- It should have sufficient mechanical strength.
- Thermal expansion should be small.
- It should be fire proof.
- It must be in pure form and not have any defects.

2.5 Ferroelectric Materials

- Ferroelectric materials exhibits spontaneous polarization i.e., they are polarized even in the absence of electric field.
- They exhibit dielectric hysteresis. The lagging of polarization behind the applied electric field is called dielectric hysteresis.
- Above ferroelectric Curie temperature, the ferroelectric material becomes paraelectric material.

- They posses ferroelectric domain structure.
- They can be polarized even by very weak electric field.
- Ferroelectric materials exhibit piezoelectricity and pyro electricity. Piezoelectricity means the creation of electric polarization by mechanical stress. Pyro electricity means the creation of electric polarization by thermal stress.

Hysteresis of Ferroelectric Materials

The ferroelectrics are known as non-linear dielectrics. Such materials exhibit hysteresis curve similar to that of ferromagnetic materials.

The lagging of polarization 'P' behind the applied electric field E is called dielectric hysteresis.

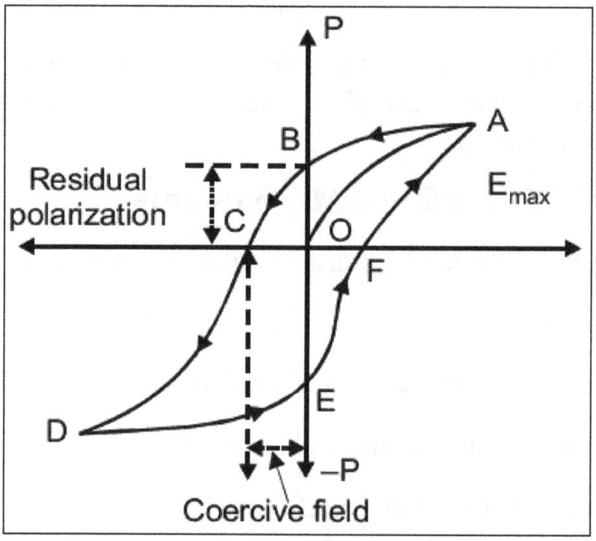

Dielectric Hysteresis in ferroelectric materials.

When a ferroelectric material is subjected to external electric field (E), its polarization (P) increases with respect to the field applied and it reaches the maximum value 'OA'.

If the applied electric field is reduced, the polarization also decreases from A and when E becomes zero, a small amount of polarization exists in the material is called spontaneous or residual polarization.

In order to reduce the value of polarization to zero, a reversing electric field OC should be applied. This field is known as coercive field.

Thus, the variation of P with respect to E traced along the closed path ABCDEFA in one full cycle of polarization and depolarization is called hysteresis or the hysteresis curve.

Application of Ferroelectric Materials

- Ferroelectric materials are used to make pressure transducers, ultrasonic transducers, microphones and gas filters.
- They are used as memory cores in computers.
- They are used to measure and control temperature.
- Ferroelectric ceramics are used as capacitors to store electrical energy.
- They are used to make very good infrared detectors.
- Rochelle salt is used in devices like microphones, strain gauges, phonograph pickups and SONAR devices.
- In optical communication, the ferroelectric crystals are used for optical modulation.
- Ferroelectric materials are used to produce ultrasonic.
- Electrets are also used to bond the fractured bones in human body.
- They are used as frequency stabilizers and crystal controlled oscillators.

2.5.1 Piezo Electrics, Pyro Electrics and Ferroelectrics

Ferro electricity refers to the creation of large induced dipole moment in a weak electric field as well as the existence of electric polarization even in the absence of an applied electric field.

Crystalline dielectric materials which possess a permanent electric polarization are called ferroelectric materials which have electric dipole moment even in the absence of any field. Normally they are anisotropic crystals which exhibit spontaneous polarization.

Examples:

- Barium Titanate (BaTiO3).
- Potassium Dihydrogen Phosphate (KDP).
- Ammonium Dihydrogen Phosphate ($NH_4H_2PO_4$).
- Lithium Niobate (LiNbO3).

2.5.2 Dielectric Materials as Electrical Insulators

When dielectrics arc used as electrical insulators in electrical machines and devices. The properties which have to be considered arc their breakdown voltage and their working temperature. The voltage at which the machine is operated should be less than the breakdown voltage of the insulator. When insulators are subjected to high electric fields, due to dielectric loss, heat will be produced and the temperature of the material increases.

If the temperature becomes very high, the insulating property may get affected because the thermal energy will cause electrons to get excited to the conduction band as discussed earlier. Therefore, it is necessary to choose proper insulators for specific applications. Insulators are classified according to the temperature range in which they can be effectively used. The classification is described in the Table.

Classification	Operating Temperature Range in oC	Examples
Class Y	Upto 90	Cotton. Silk, Paper Cotton,
Class A	Upto 105	Silk and Paper impregnated with Laquer or Polyamide Resins
Class B	Upto 130	Mica. Fibre Glass, Asbestos
Class F	Upto 155	Mica, Fibre Glass. Asbestos coated with Insulating Varnish
Class H	Upto 180	Composite Inorganic materials impregnated with Silicone
Class C	> 180	Quartz Ceramics. Teflon

2.6 Magnetic Materials: Concept of Magnetism

Magnetic permeability or simply permeability is the ease with which a material can be magnetized. It is a constant of proportionality that exists between the magnetic induction and magnetic field intensity. This constant is approximately equal to 1.257×10^{-6} Henry per meter (H/m) in free space (a vacuum). In other materials it can be much different, often substantially greater than the free-space value, which is symbolized as μo.

Materials that cause the lines of flux to move farther apart, resulting in a decrease in magnetic flux density compared with a vacuum, are called diamagnetic. Materials that concentrate magnetic flux by a factor of more than one but less than or equal to ten are termed as paramagnetic. The materials that concentrate the flux by a factor of more than ten are known as ferromagnetic. The permeability factors of some substances change with rising or falling temperature or with the intensity of the applied magnetic field.

In engineering applications, permeability is often expressed in relative, rather than in absolute, terms. If μo represents the permeability of free space (i.e., 4p X10-7H/m or 1.257 x 10-6 H/m) and μ represents the permeability of the substance in question (also specified in henrys per meter), then the relative permeability, μr is given by,

$$\mu_r = \mu / \mu_o$$

For non-ferrous metals such as copper, brass, aluminum etc., the permeability is the same as that of "free space", i.e. the relative permeability is one.

For ferrous metals however the value of μr may be several hundred. Certain ferromagnetic materials, especially powdered or laminated iron, steel or nickel alloys.

One, but no known substance has relative permeability much less than one. In addition, permeability can vary greatly within a metal part due to localized stresses, heating effects, etc.

When the paramagnetic or ferromagnetic core is inserted into a coil, the inductance is multiplied by μr compared with the inductance of the same coil with an air core. This effect is useful in the design of transformers and eddy current probes.

2.6.1 Diamagnetic, Paramagnetic, Ferromagnetic Materials

Diamagnetic

Diamagnetism is exhibited by all materials. When a material is placed in a magnetic field, the electrons in the atomic orbitals tend to counteract the external magnetic field by moving in such a way as to acquire an induced magnetic moment, which would oppose the external magnetic field. The direction of the induced dipole moment is opposite to that of the external field. Due to this effect the material gets very weakly repelled in the magnetic field.

This effect is known as diamagnetism. The induced dipole moments disappear when the field is removed. The diamagnetic effect in a material may be observed only if the paramagnetic or the ferromagnetic effect does not mask the weak diamagnetic effect. Diamagnetism may be understood by the schematic illustrations in figure (a) and (b). When the magnetic field is zero, the atoms possess zero magnetic moment figure (a) when a magnetic field Ho is applied in the direction shown figure (b). The atoms acquire an induced magnetic moment in the direction opposite to that of the field. The susceptibility of a diamagnetic material is negative. Due to this, the material gets weakly repelled in the magnetic field. The magnetization becomes zero once the magnetic field is removed. Examples of some elements that are diamagnetic are copper, gold, mercury, silver, zinc.

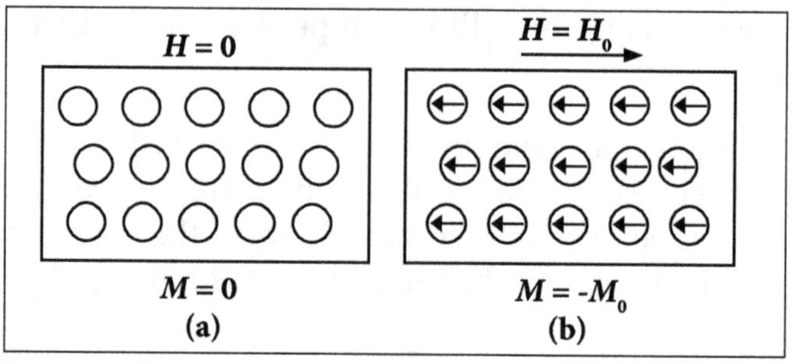

(a) The atoms do not possess magnetic moment when H=O: so M=o.
(b) When a magnetic field Ho is applied, the atoms acquire induced magnetic moment in the direction opposite to the applied field resulting in negative susceptibility.

Paramagnetic

In certain materials, each atom or molecule possesses permanent magnetic moment. In the absence of an external magnetic field, the atomic magnetic moments are all randomly oriented. This makes the net magnetic moment and hence the magnetization of the material zero. But when an external magnetic field is applied, the individual atomic magnetic moments tend to align themselves in the direction of the magnetic field resulting in a non-zero magnetization. This is illustrated in figure (a) and (b).

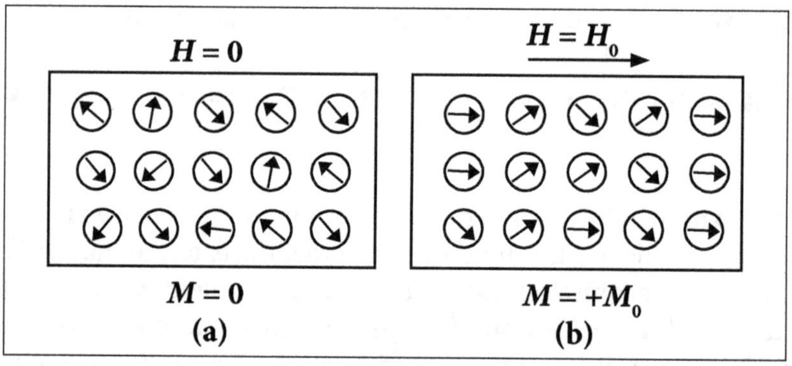

(a) Each atom possesses a permanent magnetic moment. When H=o. all the magnetic moments ore randomly oriented; so M=o. (b) When a magnetic field Ho is applied. The magnetic moments tend to orient themselves in the direction of the field, resulting in a net magnetization M = Mo and positive susceptibility.

Thermal agitation disturbs the alignment of the magnetic moments and so at higher temperatures, greater fields are required to attain the same magnetization. This implies that the paramagnetic susceptibility decreases with increasing temperature. It has been observed that the paramagnetic susceptibility varies inversely with temperature. The temperature dependence of paramagnetic susceptibility is described by the relation,

$$\chi = \frac{C}{T}$$

This is called the Curie Law of paramagnetism and C is called the Curie constant. Examples of some elements that are paramagnetic are aluminum, chromium, sodium, titanium, and zirconium.

Ferromagnetic Materials

Ferromagnetic materials have a large, positive susceptibility to an external magnetic field. They exhibit a strong attraction to magnetic fields and are able to retain their magnetic properties after an external field has been removed. Ferromagnetic materials have some unpaired electrons so their atoms have a net magnetic moment. They get their strong magnetic properties due to the presence of magnetic domains. In these domains, a large numbers of atom's moments (1012 to 1015) are aligned parallel so that the magnetic force within the domain is strong.

When a ferromagnetic material is in the unmagnitized state, the domains are nearly randomly organized and the net magnetic field for the part as a whole is zero. When a magnetizing force is applied, the domains become aligned to produce a strong magnetic field within the part. Nickel, Iron and cobalt are examples of ferromagnetic materials. Components with these materials are commonly inspected using the magnetic particle method.

Different between Diamagnetic, Paramagnetic, Ferromagnetic materials

S. No	Properties	Diamagnetic	Paramagnetic	Ferromagnetic
1.	Definition	It is a material in which there is no permanent magnetic moment.	It has permanent magnetic moment.	It has enormous (more) permanent magnetic moment.
2.	Spin or magnetic moment or dipole alignment.	No spin alignment.	Random alignment.	Parallel and orderly alignment.
3.	Behavior	Repulsion of magnetic lines of force from the center of the material.	Attraction of magnetic lines towards the center.	Heavy attraction of lines of force towards the center.
4.	Magnetized direction	Opposite to the external magnetic field.	Same direction as the external magnetic field.	Same direction as the external magnetic field.
5.	Permeability	It is very less.	It is high.	It is very high.
6.	Relativity permeability	$\mu r < 1$	$\mu r > 1$	$\mu r >> 1$

7.	Susceptibility	Negative	Low positive	High positive
8.	Magnetic phase transition	At 0 K, diamagnetic material is Superconductor. When we increase its temperature it becomes a normal conductor.	When temperature is less than the curie temp, it is converted into Diamagnetic.	When temperature of the material is greater than the Curie temperature it is converted into Paramagnet.

2.6.2 Hysteresis

Hysteresis means "Lagging" i.e., The Lagging of intensity of magnetization (I) behind the intensity of magnetic field (H).

Experimental Determination

A graph is drawn between the intensity of magnetization [I] and the intensity of magnetic field [H], for a cycle of magnetization. The experimental setup consists of solenoid coil through which current is passed and the material is magnetized.

By varying the value of current we can get different values of Intensity of magnetization [I] due to the magnetic field (H) in the solenoid.

Here,

$H \propto i,$

$H = nil$

Where,

n - No. of turns in the coil.

l - Length of the coil.

i - Current through the circuit.

In the graph,

$OB = B_r$ = Retentivity

$OC = H_C$ = Coercivity

1. When the intensity of magnetic field 'H' is increased from O to F, the value of Intensity of magnetization also increases from O to A, at 'A' the material reaches the saturation value of Intensity of magnetization. Then the value of I is constant.

2. When intensity of magnetic field 'H' is decreased from G to O, the value of Intensity of magnetization 'I' also decreases from A to B, but not to zero (0). Now the material

retains [stores] some amount of magnetism known as Retentivity, even though the intensity of magnetic field 'H' is zero. It is represented as 'OB' in the graph.

3. When intensity of magnetic field 'H' is increased in reverse direction from O to C, the value of Intensity of magnetization 'I' decreases from B to C. i.e., the value of 'I' reaches zero.

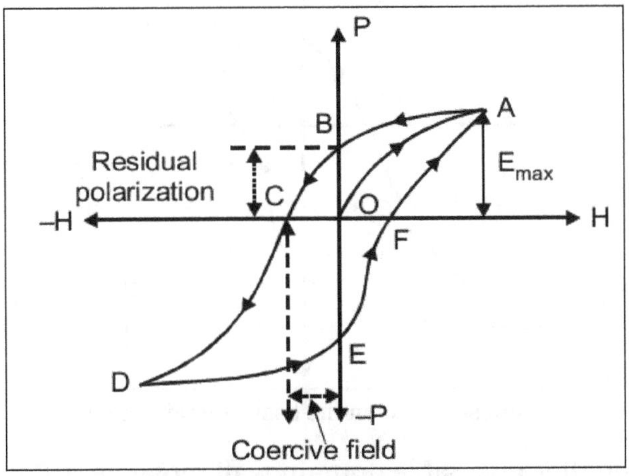

Hysteresis curve.

4. The amount of intensity of magnetic field 'H' applied in the reverse direction to remove the retentivity is known as Coercivity or Coercive force. It is represented as 'OC' in the graph.

5. Further repeating the process, the remaining portion [CDEFA] in the graph is obtained. The closed loop [OABCDEFA] is called Hysteresis loop or (I - H) curve for one cycle of magnetization.

6. Now the material is taken out. After a cycle of magnetization, there is some expenditure (loss) of energy.

7. This loss of energy is radiated in the form of heat energy in the material.

8. This loss of energy is directly proportional to the area of the loop.

9. From the Hysteresis graph, we can select soft and hard magnetic materials depending upon the purpose.

Energy Product

It is the product of residual magnetism Br and coercivity, which HC gives the maximum amount of energy stored in the specimen.

$$\text{Energy product} = B_r \times H_C$$

Hysteresis Loss

When the specimen is taken through a cycle of magnetization, there is a loss of energy in the form of heat. This loss of energy is known as Hysteresis loss.

Explanation of Hysteresis on the Basis Of Domain Theory

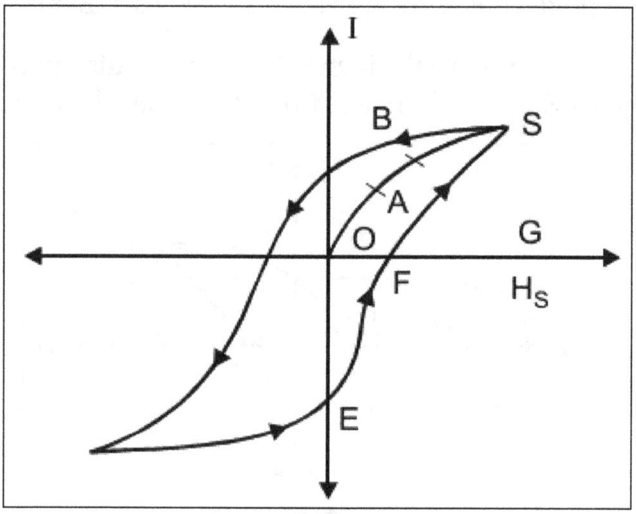

Hysteresis curve on the basis domain theory.

OA - Due to smaller reversible domain wall movement.

AB - Due to larger irreversible domain wall movement.

BS - Due to smaller irreversible domain rotation.

S - Point of saturation.

1. When a field is applied, for small H, the domain walls are displaced and gives rise to small value of magnetization. [OA in the graph]. Now, the field is removed, the domains return to its original state known as reversible domains.

2. When the field increases, a large number of domains contribute to the magnetization and I increases rapidly with H.

3. Now, when the field is removed the domain boundaries do not come back to the original position due to the domain wall movement to a very large distance (AB in the graph). These domains are called irreversible domains.

4. Now if the field is further increased, domains start rotating along the field direction and anisotropic energy is stored and it is represented as BC in the graph.

5. Thus the specimen is said to attain maximum magnetization at this position even after the removal of the field. The material has magnetism called retentivity. This retentivity can be destroyed by applying a high reverse magnetic field called coercivity.

6. Thus the reversible and irreversible domain wall movements give rise to hysteresis in the Ferromagnetic materials.

2.6.3 Soft & Hard Magnetic Materials

Hard Magnetic Materials

The materials which are very difficult to magnetize and demagnetize are called hard magnetic materials. These materials can be made by heating the magnetic materials and then cooling it suddenly. It can also be prepared by adding impurities.

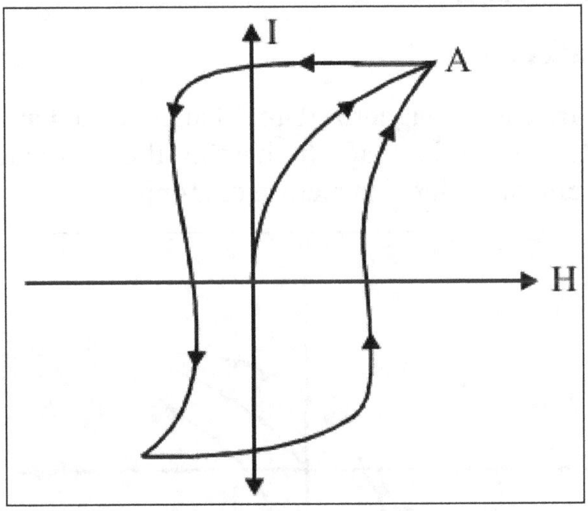

Hysteresis loop of a hard magnetic material.

The above hysteresis loop is very hard and has a large loop area for a hard magnetic material, therefore the loss is also large.

Domain wall does not move easily and require large value of H for magnetization. Its coercivity and retentivity values are large. Its eddy current loss is also high due to its low resistivity, the magneto static energy is large. It has low susceptibility and permeability. The hard magnetic materials have large amount of impurities and lattice defects.

Examples: Tungsten steel, Carbon steel, Chromium steel, Alnico etc.

Properties

- It is very hard to magnetize and also demagnetize.
- The hysteresis curve is very broad.
- The coercivity and retentivity values are large.
- These materials have low value of susceptibility and permeability.
- The magneto static energy is large.
- The eddy current loss is very high.

Applications

- Magnets made by carbon steel are used for manufacturing the toys and compass needle.
- Tungsten steel is used in making permanent magnets for D.C motors.
- It is also used for making a small size of magnets.

Soft Magnetic Materials

The materials which are easily magnetized and demagnetized are called soft magnetic materials. These materials can be made by heating the magnetic materials and then cooling it slowly to attain an ordered structure of atoms.

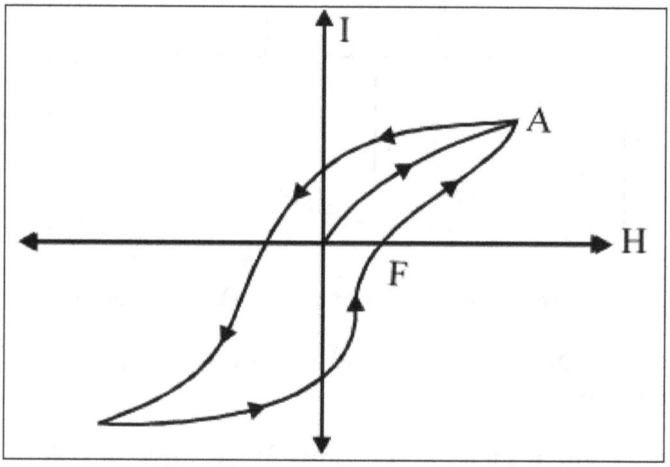

Hysteresis loop of a soft magnetic material.

The above hysteresis loop is very small and has a less loop area for a soft magnetic material. Therefore the loss is also small. Domain wall move easily and require small value of H for magnetization. Its coercivity and retentivity values are small its eddy current loss is small due to its high resistivity.

The magneto static energy is less, it has high value of susceptibility and permeability. The soft magnetic materials do not have impurities and lattice defects.

Examples: Iron-Silicon alloys, Nickel-Iron alloys and Iron-cobalt alloys.

Properties

- It is easily magnetized and demagnetized.
- They hysteresis area is very small and hence, the hysteresis loss is also small, as shown in figure.
- The coercivity and retentivity are very small.

- These materials have large values of susceptibility and permeability.
- Their magneto static energy is very small.
- The eddy current loss is very small.

Applications

- Iron-Silicon alloys are used in electrical equipment and magnetic cores of transformers.
- Cast iron is used in the structure of electrical machinery and the frame work of D.C machine.
- Nickel alloys are used to manufacture inductors, relays and small motors.
- It is also used for computer and data storage devices.

2.6.4 Ferrite

Ferrites or Ferromagnetic materials are the modified structure of iron without carbon. In Ferrites, the spins of adjacent ion in the presence of a magnetic field are in opposite directions with different magnitudes.

Properties

- These are made from ceramic ferromagnetic compounds.
- It has low tensile strength and it is brittle and soft.
- In these materials, all the valence electrons are tied up by ionic bonding.
- These are bad conductors with high resistivity of the order of 1011m.
- Ferrites have low eddy current loss and low hysteresis loss.
- The general formula for Ferrites is $X^2+ (Fe_2)^3+ O_4$ where X may be a metal such as Mg, Ni, Mn, Zn, etc.
- Ferrites are manufactured by powder metallurgical process by mixing, compacting and then sintering at high temperatures followed by age hardening in magnetic fields.

Structures of Ferrites

Ferrites are the magnetic compounds consisting of two or more different kinds of atoms. Generally, ferrites are expressed as $X^2+(Fe_2)^3+O_4$ where X^{2+} stands for suitable divalent metals ions such etc Mg^{2+}, Zn^{2+}, Fe^{2+}, Mn^{2+}, Ni^{2+} etc.

Normally, there are two types of structures present in the ferrites:

- Regular spinel.
- Inverse spinel.

Regular Spinal

In the regular spinal type, each metal atom (divalent) is surrounded by four ions in a tetragonal fashion.

For example in Mg^{2+}, Fe_{23+}, O_4^{2+}, the structure of Mg^{2+} is given in the figure and it is called “A’ site. Totally in a unit cell, there will be 8 tetrahedral (8A) sites.

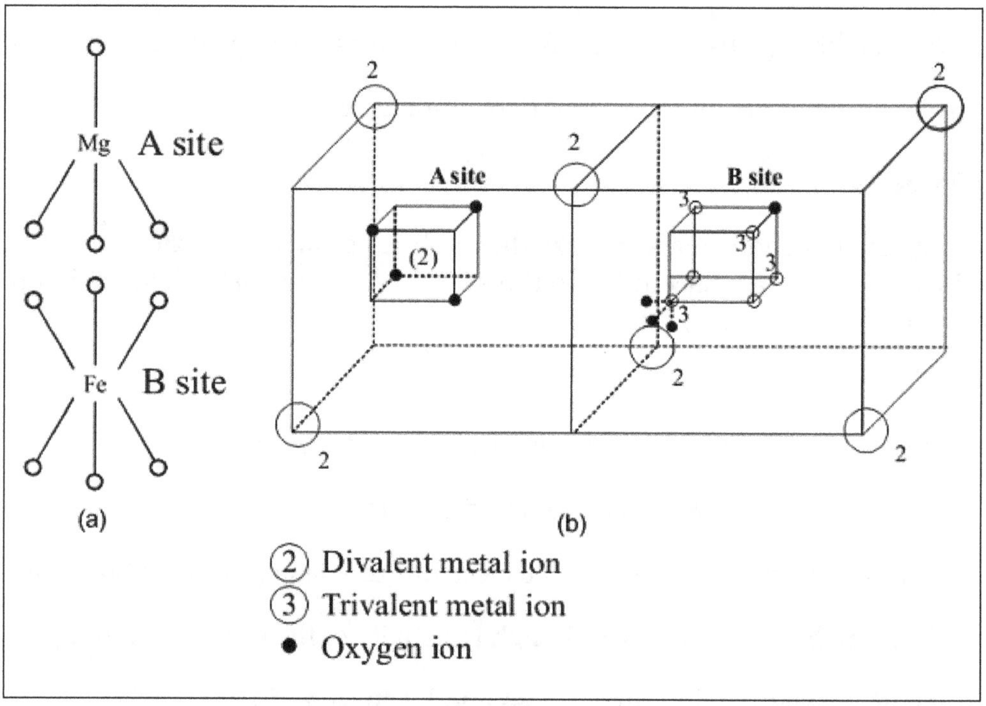

Structure of ferrites (Regular Spinal).

Thus in a regular spinal, each divalent metal ion (mg^{2+}) exists in a tetrahedral form (A site) and each trivalent metal ion (Fe^{2+}) exists in an octahedral form (B site).

Hence, the sites A and B combine together to form a regular spinal ferrite structure as shown in figure.

Inverse Spinal

In this type, we consider the arrangement of dipoles of a single ferrous ferrite molecule Fe^{3+} [Fe^{2+} Fe^{3+}] O_4^{2-} Fe^{3+} ions (trivalent) occupies all A sites (tetrahedral) and half of the B sites (octahedral) also. Thus, the left out B sites will be occupied by the divalent (Fe2+). The inverse spinal structure is shown in the figure.

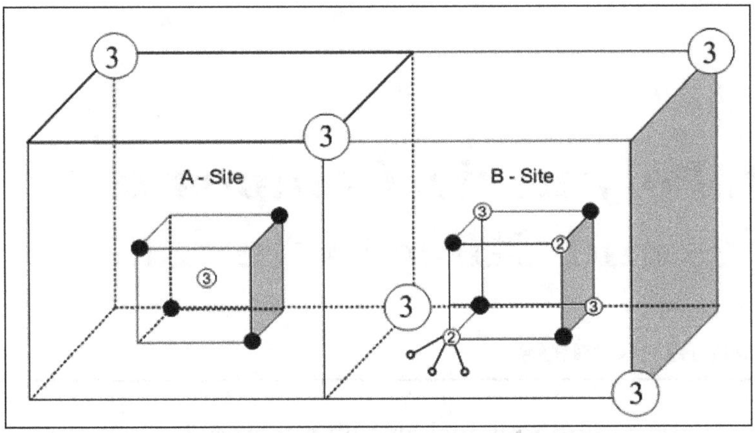

Structure of ferrites (inverse Spinal).

Application of Ferrites

- Ferrites are used to produce ultrasonic wave by Magnetostriction principle.
- Ferrites are used in audio and video transformers.
- Ferrites rods are used in radio receivers to increase the sensitivity.
- Ferrites are widely used in non-reciprocal microwave devices such as gyrator, circulator and Isolator.
    - Gyrator: It transmits the power freely in both the directions with a phase shift of radians.
    - Circulator: It provides sequential transmission of power between the ports.
    - Isolator: It is used to display differential attenuation.
- They are also used for power limiting and harmonic generation.
- Ferrites are used in parametric amplifiers so that the input can be amplified with low noise.
- They are used in computers and data processing circuits.
- Bi-stable elements, Ferro cube (Ferrite with square hysteresis loop), magnetic shift register and magnetic bubbles are also examples for Ferrites.

3

Optical, Polymeric, Composite, Ceramics and Other Materials

3.1 Optical Materials

When electromagnetic radiation is incident on a material, the material responds in a characteristic manner which depends on the type of interaction between the electromagnetic radiation and the electrons of the atoms in the material. Optical characteristics of materials are determined by the interaction of the electrons with the radiation which lies within or near visible range of the electromagnetic spectrum.

Electromagnetic radiation prevails everywhere in the form of light (visible radiation) and it is incident on all the objects around us all the time. As our eyes are sensitive detectors of light radiation, we detect or see many of the common optical characteristics of materials such as their colour, brightness, transparency, opacity, reflectivity, etc. Besides these common properties, there are many more special optical properties of materials which make them useful in a wide range of optical devices. Some of the commonly used optical materials and devices in our daily use are, window glasses, lenses, mirrors, anti-reflection coatings, etc.

Some of the more recently developed high-technology optical devices are lasers, optical fibres, photodiodes, optical memories (CD-ROM), electro-optic modulators, etc.

When light falls on a material any one or more of the following phenomena may occur:

- Scattering.
- Refraction.
- Reflection.
- Transmission.
- Absorption.
- Luminescence.

All these phenomena can be understood in terms of the interaction of the incident light (electromagnetic radiation) with the electrons of the material.

3.1.1 Scattering, Refraction, Reflection, Transmission and Absorption

Scattering

When electromagnetic radiation is incident on a material, the oscillating electric field of the incident radiation forces the electrons in the material to oscillate at the frequency of the incident radiation. An oscillating charge acts as a source of electromagnetic radiation. As the electrons oscillate, they produce radiation of the same frequency as that of the incident radiation. The radiation emanating from the oscillating electrons which travels in all directions is called the scattered radiation. Almost all objects scatter radiation and this is what makes the object visible to us.

Refraction

The radiation that is incident on a material gets scattered out of the material. The newly generated radiation, due to the oscillating dipoles interferes with the incident radiation and propagates into the material with phase retardation. The speed of the wave that propagates inside the material is lower than that of the incident radiation. The speed depends on the number of electrons that are en-countered per unit length and their polarizability. The speed is said to depend on the optical density of the material. At optical frequencies, the polarization that is effective is only the electronic polarization.

A characteristic parameter called the refractive index is introduced which is a measure of the change in speed of the wave. Refractive index n is defined as the ratio of the speed of the electromagnetic wave in vacuum (or air) Vvac to its speed in the material Vmat,

$$n = \frac{\upsilon_{vac}}{\upsilon_{mat}} = \frac{c}{\upsilon_{mat}}$$

$$c = 3 \times 10^8 \text{ m/sec.}$$

Where the refractive index n defined above is the absolute refractive index because it is defined with respect to vacuum. If the radiation travels from medium 1 of absolute refractive index n_1 into another medium 2 of absolute refractive index n_2, then n_{21} is called the relative refractive index (of medium 2 with respect to medium 1). n_{21} will be greater than one if medium 2 is optically denser than medium 1 and it will be less than one if medium 1 is optically denser than medium 2,

$$\frac{n_2}{n_1} = \frac{\upsilon_1}{\upsilon_2} = n_{21}$$

The refracted ray, as it passes through the material may get partially absorbed by the material and the rest may get transmitted.

Reflection

In general, reflection, transmission and absorption depend on the wavelength of the affected radiation. Thus, these three processes can either be quantified for monochromatic radiation or for a certain kind of polychromatic radiation. For the latter, the spectral distribution of the incident radiation has to be specified. In addition, transmittance, reflectance and absorptance might also depend on polarization and the geometric distribution of the incident radiation, which therefore also have to be specified.

The reflectance ρ is defined by the ratio of reflected radiant power to incident radiant power. For a certain area element dA of the reflecting surface, the incident radiant power is given by the surface's irradiance Ee, multiplied with the size of the surface element,

Thus,

$$dF_{e,incident} = E_e \; dA$$

And the (differential) reflected radiant power is given by the exitance M_e, multiplied with the size of the surface element,

$$dF_{e,reflected} = M_e \; dA$$

Thus,

$$\rho = \frac{d\Phi_{e,reflected}}{d\Phi_{e,incident}} = \frac{M_e \cdot dA}{E_e \cdot dA} = \frac{M_e}{E_e}$$

or,

$$M_e = r \, E_e$$

Total reflectance is further subdivided in regular reflectance rr and diffuse reflectance ρr, which are given by the ratios of regularly (or specularly) reflected radiant power and diffusely reflected radiant power to incident radiant power. From this definition, it is obvious that r = rr + rd.

Transmission and Absorption

The transmittance t of a medium is defined by the ratio of transmitted radiant power to incident radiant power. Total transmittance is further subdivided in regular transmittance t_r and diffuse transmittance t_d, which are given by the ratios of regularly (or directly) transmitted radiant power and diffusely transmitted radiant power to incident radiant power.

Again,

$$t = t_r + t_d$$

The absorptance a of a medium is defined by the ratio of absorbed radiant power to incident radiant power. Being defined as ratios of radiant power values, reflectance, transmittance and absorptance are dimensionless. Quantities such as reflectance and transmittance are used to describe the optical properties of materials. The quantities can apply to either complex radiation or to monochromatic radiation. The optical properties of materials are not a constant since they are dependent on many parameters such as:

- Thickness of the sample.
- Angle of incidence.
- Surface conditions.
- Temperature.
- The spectral composition of the radiation (CIE standard illuminants A, B, C, D65 and other illuminants D).
- Polarization effects.

That part of the incident radiation, that is not reflected will get partly transmitted and partly absorbed. The transmittivity is defined as the ratio of the intensity of light transmitted through the material to the intensity of light that is incident on the material.

If there is no absorption,

$$T = \frac{I_T}{I_o}$$

$$R + T = 1$$

3.2 Laser: Principles and Applications

A laser is a coherent and focused beam of photons. Coherent means that it is all one wavelength, unlike ordinary light which showers on us in many wavelengths.

Laser is the acronym for light amplification by stimulated emission of radiation. Ideal laser light has single - wavelength only. This is related to the molecular characteristics of the material being used in the laser. It is formed in parallel beams and in single phase. That is, it is coherent.

The losing medium can be a gas, a liquid, an insulating crystal (solid state) or a semiconductor. For optical fiber systems, semiconductor laser diodes are used as a optical source. Because its output radiation is highly monochromatic and the light beam is very directional.

Semiconductor laser diodes are preferred over LED for the optical fiber communication systems requiring bandwidth greater than approximately 200 MHz.

Characteristics of Laser Diode

Laser diodes have the following characteristics:

- Response time less than 1 ns.
- Optical bandwidth of 2 nm or less.
- High coupling efficiency.
- Laser diodes multilayered.
- Smaller temp dependence.
- Coherent sources.

Coherent light is the light in which the photons are all in 'step'. In other words, the change of phase within the beam occurs for all the photons at the same time. There are no abrupt phase changes within the beam. Light produced by lasers is both coherent and monochromatic.

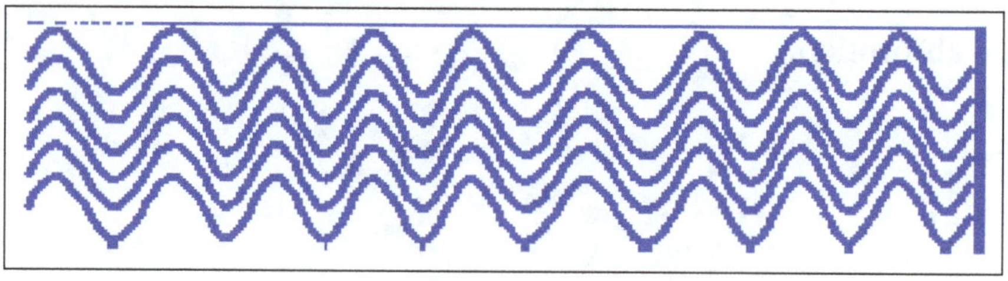

Coherent source.

Characteristics of Lasers

1. Laser Light is Highly Coherent

Wave trains for laser light may be several hundred kilometers long. Interference fringes can be set up by combining two beams that have followed separate paths whose lengths differ by as much as this amount. The corresponding coherence length for light from a tungsten filament lamp or a gas discharge tube is considerably less than 1 m.

2. Laser Light is Highly Monochromatic

The tungsten light, spread over a continuous spectrum, gives us no basis for comparison. The light from selected lines in a gas discharge tube, however, can have wavelengths in the visible region that are precise to about 1 part in 106. The sharpness of the definition of laser light can easily be a thousand times greater or 1 part in 10_9.

3. Laser Light can be Sharply Focused

This property is related to the parallelism of the laser beam. As for star light, the size of the focused spot for a laser beam is limited only by the diffraction effects and not by the size of the source. Flux densities for focused laser light of 1015Wcm-2 are readily achieved. An oxyacetylene flame has a flux density of only 103Wcm-2.

4. Tuning

Some lasers can be used to emit radiation over a range of wavelengths. Laser tunability leads to applications in photochemistry, high resolution and roman spectroscopy.

5. Brightness

The primary characteristics of the laser radiation is that lasers have a higher brightness than any other light source. We define the brightness as the power emitted per unit area per unit solid angle.

Population Inversion: Three and Four Level Pumping Schemes

Stimulated emission process is a key factor for the production of laser. For stimulated emission, more number of atoms must be in excited state. Establishing the situation in which no. of atoms in higher energy state is more than that in lower energy state is called population inversion. Hence, the population inversion is needed in the production of laser.

When a sizable population of electrons resides in upper levels, this condition is known as "population inversion" and it sets the stage for stimulated emission of multiple photons. This is the precondition for light amplification which occurs in a LASER and since the emitted photons have a definite time and phase relation to each other, the light has a high degree of coherence.

Three and Four Level Pumping Schemes

In a simple two-level system, it is not possible to obtain a population inversion with optical pumping because the system can absorb pump light only as long as population inversion and light amplification is not achieved. Essentially, the problem is stimulated emission that is caused by the pump light itself.

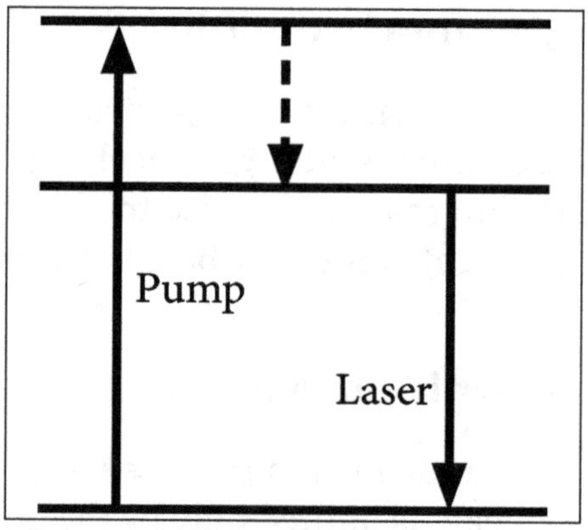

Three-level system.

Inversion by optical pumping becomes possible when using a three-level system. Pump light with a shorter wavelength can transfer atoms from the ground state to the highest level. From there, spontaneous emission or a non-radiative process (e.g., involving phonons in a laser crystal) transfers atoms to an intermediate level, called the upper laser level. From that level down to the ground state, the laser transition with stimulated emission can occur. With sufficiently high pump intensity, population inversion for the laser transition can be reached as the stimulated emission by the pump radiation is prevented by the transferring to the intermediate level.

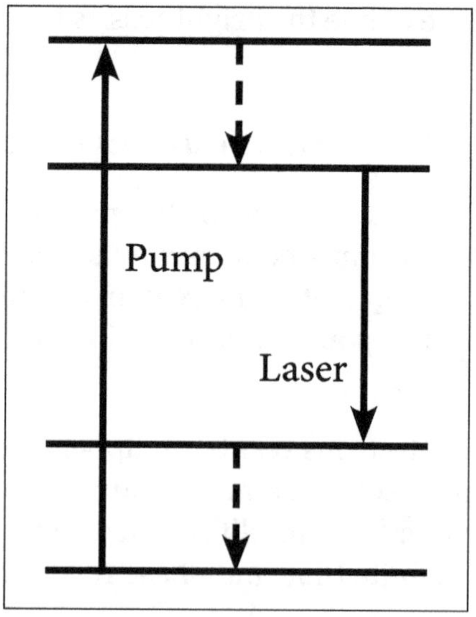

Four-level system.

Laser gain with a much lower excitation level is possible in a four-level system, such as Nd: YAG. Here, the lower level of the laser transition is somewhat above the ground state and a rapid. Transfer from there to the ground state keeps the population of the lower laser level very small. Therefore, moderate population in the third level (the upper laser level), as achieved with a moderate pump intensity, is sufficient for laser amplification.

Applications of Laser

There are many scientific, industrial, military, medical and commercial laser applications. The coherency, high mono chromaticity, focus ability, directionality and ability to reach extremely high powers are all properties which allow for these specialized applications.

Application of lasers in medicine:

- Laser are used for micro surgery and bloodless operations to cure cancer and skin tumors in human beings and animals.
- They are used for the treatment of detached retina.
- Lasers are also used for removal of tattoos or any superficial unwanted growth.

Application of Lasers in Industry

Laser Cutting

In industry lasers are used for the precise cutting of flat materials. Lasers have the advantage that there is no physical contact with the material so there is no chance of contamination, also there is less chance of the material warping as the laser energy can be focused on a very small area so the whole material is not heated. Even a three dimensional profile can be cut using lasers. Laser cutting is also employed in tailoring industry.

Laser Welding

In laser welding, a beam of laser is focused on to the spot to be welded. Due to the heat generated, the material melts over a tiny area and upon cooling the material becomes homogeneous solid structure. Laser welding is a contact-less process and thus no outside material gets into the welded region. Since the heat affected zones are very small, laser welding is ideal for many microelectronic devices.

Laser Drilling

Laser drilling of holes is achieved by subjecting the material to powerful laser pulses of about millisecond duration. The intense heat generated over a short duration by the pulses evaporates the material locally leaving a hole. Very fine holes of the dimensions of one tenth of millimeter can be drilled. Since there is mechanical stress involved in laser drilling, even the brittle materials can be drilled.

Measurement of Pollutants in the Atmosphere

There are various types of pollutants in the atmosphere. In the measurement of pollutants, laser is used in the way a radar system is used. Hence it is called LIDAR meaning

Light Detection and Ranging. This can evaluate the distance, altitude and angular co-ordinates of the object.

3.2.1 Optical Fibres: Principles and Applications

Optical fiber communication elements operate in the micrometer wavelength zone of the frequency spectrum.

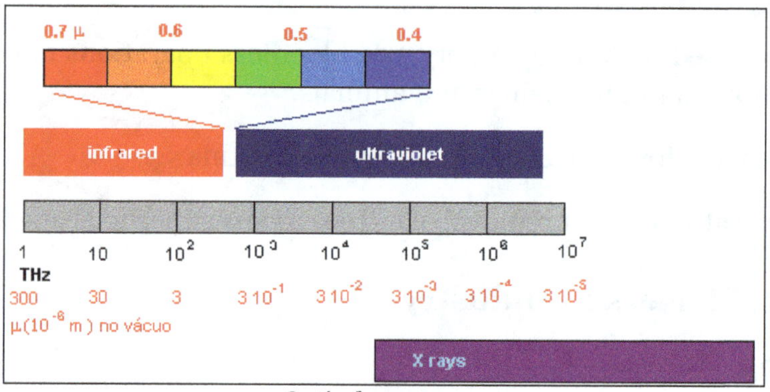

Optical spectrum.

In space light propagates with velocity c =3x108 m/s.

In a transparent medium, the speed of light is affected by a factor n, the refractive index of the medium is given by

$$n = \frac{\text{Speed of Light in Free space}}{\text{Speed of Light in the Medium}}$$

When a light ray encounters the interface of a medium, the light ray is refracted and its direction of propagation changes according to Snell's law of refraction.

$$n_1 \sin(\phi_1) = n_2 \sin(\phi_2)$$

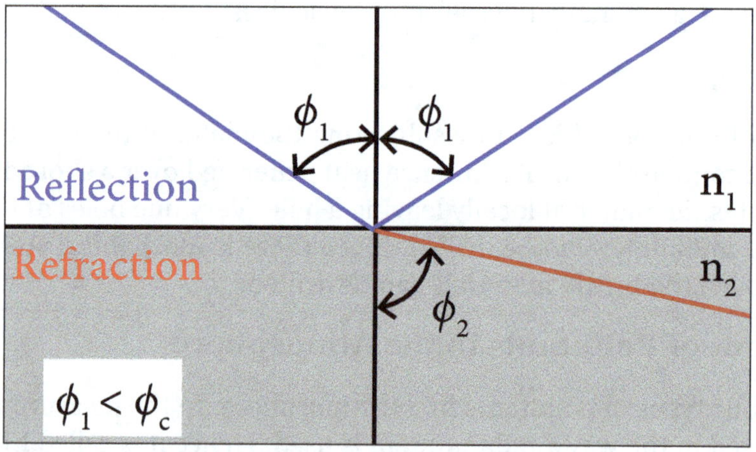

(a) Refraction-reflection of waves.

If $n_1>n_2$, at a certain angle of incidence, the refracted wave propagates parallel to the interface and the angle of refraction reaches 90° and the corresponding angle of incidence is called the critical angle.

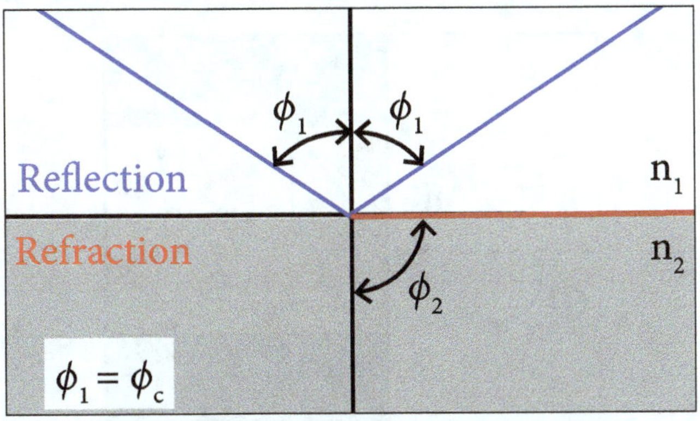

(b) Refraction-reflection theory (critical angle).

For angles of incidence greater than the critical angle, the rays are totally reflected. Total internal reflection occurs.

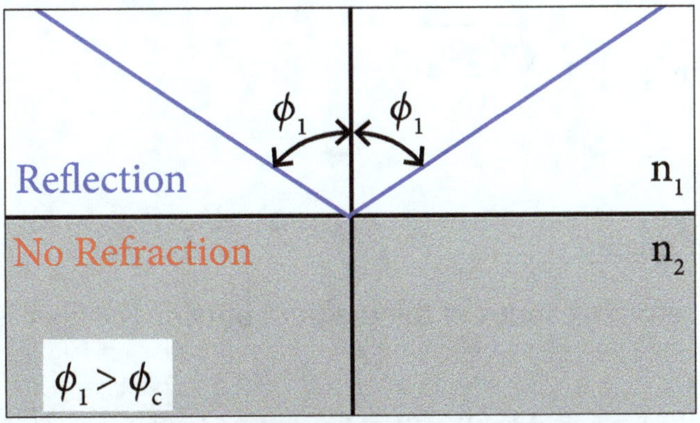

Refraction-reflection theory (total reflection).

Optical fibers are cylindrical waveguides made of two concentric layers of very pure glass. The core (the interior layer) with refractive index n_1 serves as the medium for light propagation, while the cladding (the exterior layer) has a lower refractive index n_2, where $n_1>n_2$ assuring that light rays are reflected back to the core.

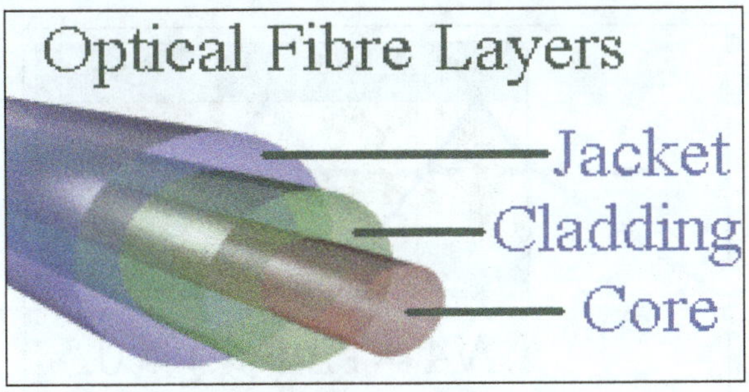

Optical fiber structure.

Acceptance Angle and Acceptance Cone-Numerical Aperture

Based on the principle of total internal reflection, only rays with incident angles greater than the critical angle, at the cladding-core interface can be transmitted.

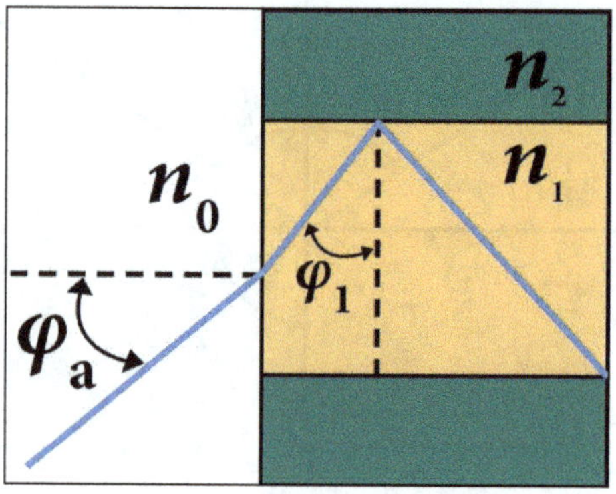

Representation of the acceptance angle.

The acceptance of light into the core is defined by the acceptance angle of the fiber, whose maximum value can be calculated directly by,

$$\sin(\phi_a) \le \sin(\phi_{aMAX}) = \frac{\sqrt{n_1^2 - n_2^2}}{n_o}$$

Where,

ϕ_a is the acceptance angle.

Assuming that the refractive index of the medium outside the fiber is 1 for air, it can be expressed as,

$$(\text{Refractive index outside the fibre})\, n_o = 1 \Rightarrow \sin(\phi_{aMAX}) = \sqrt{n_1^2 - n_2^2}$$

The numerical aperture measures the light gathering capacity of a fiber and is the sine of the maximum acceptance angle,

$$NA = \sin(\phi_{aMAX})$$

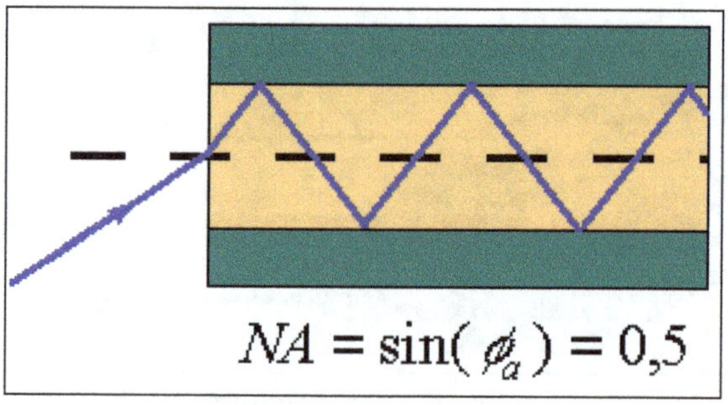

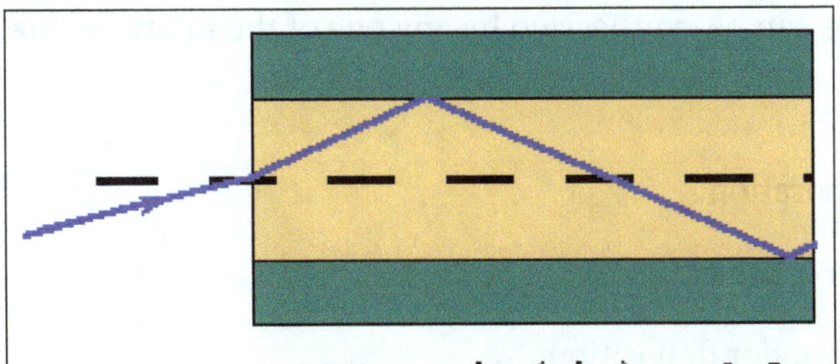

Numerical Aperture representation.

Normalized Refractive Index Difference

Another important parameter is the normalized refractive index difference which can be represented by,

$$\Delta = \frac{n_1^2 - n_2^2}{2n_1^2}$$

Whose relation with the numerical aperture is given by,

$$NA = n_1\sqrt{2\Delta}$$

The normalized frequency V is also an important quantity. This depends on the core radius, the numerical aperture of the core and the wavelength of the light stimulating the fiber,

$$V = 2\pi \times \frac{a}{\lambda} \times NA$$

The V parameter allows the determination of the fiber operating in the regime:

With only one propagating mode (mono mode propagation) or several propagating modes (multimode propagation), the last situation leads to the increased distortion of the signal due to the different propagation velocities of the modes. Reduction of material dimensions has pronounced effects on the optical properties. Optical properties of nano-materials are different from bulk forms. The change in optical properties is caused by two factors:

- The quantum confinement of electrons within the nano particles increases the energy level spacing.
- Surface plasma resonance, which is due to smaller size of nano-particles than the wavelength of incident radiation.

Carbon nanotubes can be synthesized by any one of the following methods:

- Pyrolysis.
- Laser evaporation.
- Carbons are method.
- Chemical vapour deposition.

Pyrolysis

Carbon nanotubes are synthesized by the pyrolysis of hydro carbons such as acetylene at about 700°C in the presence of Fe - silica or Fe - graphite catalyst under inert conditions.

Laser Evaporation

A quartz tube containing argon gas and graphite target is heated to 1200°C. The quartz tube containing the water – cooled copper collector which is present outside the furnace. The graphite target containing small amount of cobalt and nickel which act as a catalyst for the formation of nanotubes.

An intense pulsed laser beam is incident on the target which results in evaporation of carbon from the graphite. The argon gas carried the carbon atoms from the high temperature zone to colder copper collector, where it condense into nano tubes.

Carbon Arc Method

It is carried out by applying direct current 60 – 100A and 20 – 25V are between graphite electrodes of 10 – 20μm diameter.

Chemical Vapour Deposition

It involves the decomposition of hydrocarbon gas (methane gas) at 1100°C. As the gas decomposition, carbon atoms are produced. Then it is condensed on a cooler substrate which contains catalyst like iron. This method produced tubes with open end which allows continuous fabrication. It is the most favorable method for scale up and production.

Properties of Nano-Materials

1. Melting Points

Nano-materials have a significantly lower melting point and appreciable reduced lattice constants. This is due to huge fraction of surface atoms in the total amount of atoms.

2. Optical Properties

Reduction of material dimensions has pronounced effects on the optical properties.

Optical properties of nano-materials are different from bulk forms. The change in optical properties is caused by two factors:

- The quantum confinement of electrons within the nano-particles increases the energy level spacing.
- Surface plasma resonance, which is due to smaller size of nano-particles than the wavelength of incident radiation.

3. Magnetic Properties

Magnetic properties of nano-materials are different from that of bulk materials. Ferromagnetic behaviour of bulk materials disappear when the particle size is reduced and transfer to super paramagnetic. This is due to the huge surface area.

4. Mechanical Properties

The nano-materials have less defects compared to bulk materials, which increases the mechanical strength:

- Mechanical properties of polymeric materials can be increased by the addition of nano-fillers.
- As nano-materials are stronger, harder and more wear resistant and corrosion resistant, they are used in spark plugs.

Applications of Nano-Materials

Nano-materials posses unique and beneficial, physical, chemical and mechanical properties, they can be used for a wide verity of applications.

Material Technology

- Nano-crystalline aero-gel are light weight and porous, so they are used for insulation in offices homes, etc.
- Cutting tools made of nano-crystalline materials are much harder, much more wear-resistance and last stranger.
- Nano-crystalline material sensors are used for smoke detectors, ice detectors on air craft wings, etc.
- Nano-crystalline materials are used for high energy density storage batteries. Nano sized titanium dioxide and zinc dioxides are used in sunscreens to absorb and reflect ultraviolet rays.
- Nan coating of highly activated titanium dioxide acts as water repellent and

antibacterial. The hardness of metals can be predominately enhanced by using nano-particles.

- Nano-particles in paints change colour in response to change in temperature or chemical environment and reduce the infrared absorption and heat loss.
- Nano-crystalline ceramics are used in automotive industry as high strength springs, ball bearings and valve lifters.

Information Technology

- Nano-scale fabricated magnetic materials are used in data storage Nano computer chips reduces the size of the computer.
- Nano-crystalline starting light emitting phosphors are used for flat panel displays.
- Nano-particles are used for information storage.
- Nano-photonic crystals are used in chemical optical computers.

Bio-Medicals

- Bio-sensitive nano-materials are used for ragging of DNA and DNA chips.
- In the medical field, nano-materials are used for disease diagnosis, drug delivery and molecular imaging.
- Nano-crystalline silicon carbide is used for artificial heart valves due to its low weight and high strength.

Energy Storage

- Nano-particles are used hydrogen storage.
- Nano-particles are used in magnetic refrigeration.
- Metal nano-particles are useful in fabrication of ionic batteries.

3.3 Polymeric Materials: Types of Polymers

Classification of Polymers

They are classified into two types:

- Natural polymers.

- Synthetic polymers.

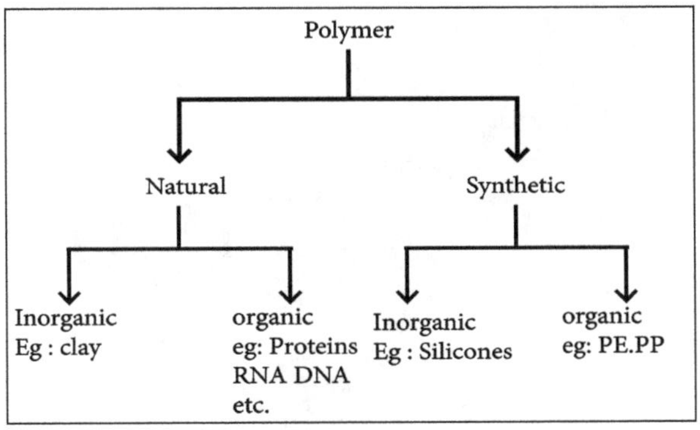

Natural Polymers

The polymers obtained from nature (plants and animals are called natural polymers).

Example:

Starch: It is a polymer of glucose. It is a chief food reserve of plants.

Cellulose: It is a polymer of glucose. It is a chief structural material of the plants.

Both Starch and cellulose are produced by plant during photosynthesis.

National rubber: It is a polymer of unsaturated hydrocarbon, 2-methyl-1, 3 butadiene, called isoprene. It is obtained from latex of rubber trees.

$$\underset{\text{Isoprene}}{nCH_2 = \underset{\underset{CH_3}{|}}{C} - CH = CH_2} \xrightarrow{\text{Polymerisation}} \underset{\text{Polyisoprene (NR)}}{\left(CH_2 - \underset{\underset{CH_3}{|}}{C} = CH - CH_2 \right)_n}$$

Synthetic Polymers

The polymers which are prepared in the laboratories are called synthetic polymers. These are also called man-made polymers.

Example: PVC, poly ethylene, Teflon, Bakelite etc.

Synthetic Polymer-types

1. Organic Polymer

These are polymers containing hydrogen, oxygen, nitrogen and sulfur and halogen atoms apart from carbon atoms.

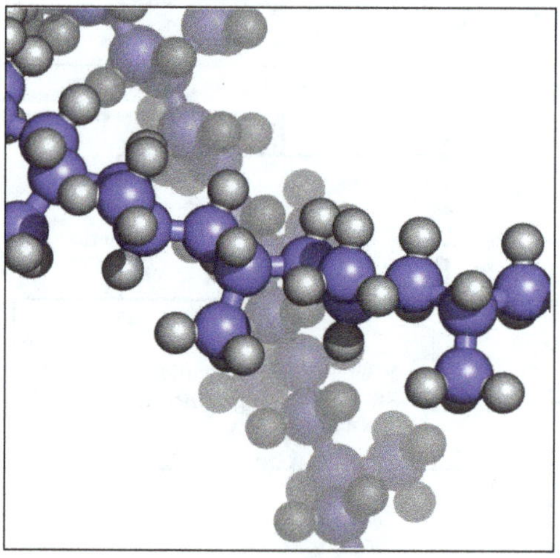

Synthetic organic polymers.

Example: PVC (Polyvinyl chloride), Epoxy polymer, polyurethane.

2. Element Organic

These are polymers composed of carbon atoms and heteroatoms (like N, S & O). The main chain consists of carbon atoms and whose side groups contain hetero atoms linked directly to the 'c' atoms in the chain.

Example: Polysiloxane.

3. Inorganic Polymers

These are polymer contain no carbon atoms. The chains of these polymers are composed of different atoms joined by chemical bonds.

Example:

(a) **Poly silicon dioxide** 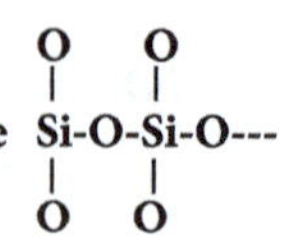

(b) **Polyphosphoric acid** 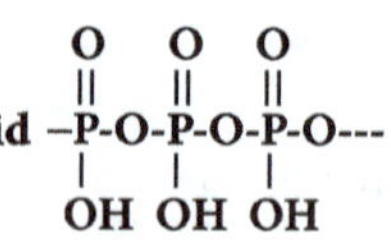

3.3.1 Mechanism of Polymerization

Polymerization can proceed according to two different mechanisms referred to as chain-growth and step-growth polymerization. In chain-growth polymerization (also called addition polymerization) reaction occurs by successive addition of monomer molecules to the reactive end (e.g. a radical end) of a growing polymer chain.

The most important group of chain-growth polymerizations is polymerization of vinyl monomers such as ethene, propene, styrene and vinyl chloride:

$$nCH_2 = CHX \rightarrow -(-CH_2 - CHX)_n - \text{in which } X = H, CH_3,\ C_5H_6 \text{ or Cl respectively.}$$

An initiator or a catalyst is usually required to start the chain-growth reaction.

Step-growth polymerization involves the reaction between the functional groups (HO-, HOOC-, etc.) of any two molecules (either monomers or polymers). By repeated reaction long chains are gradually produced.

Most commonly the reactions are condensation reactions resulting in the elimination of a small molecule like water or methanol. Examples are the production of poly (ethene terephthalate) by reaction of ethene glycol with either terephthalic acid or dimethyl terephthalate.

The principle reactions taking place when two different functional groups are present on the same molecule (indicated as AB. e.g. in the production of caprolactam) and when two different molecules each contain two identical functional groups (indicated as AA and BB, e.g. in production of PET).

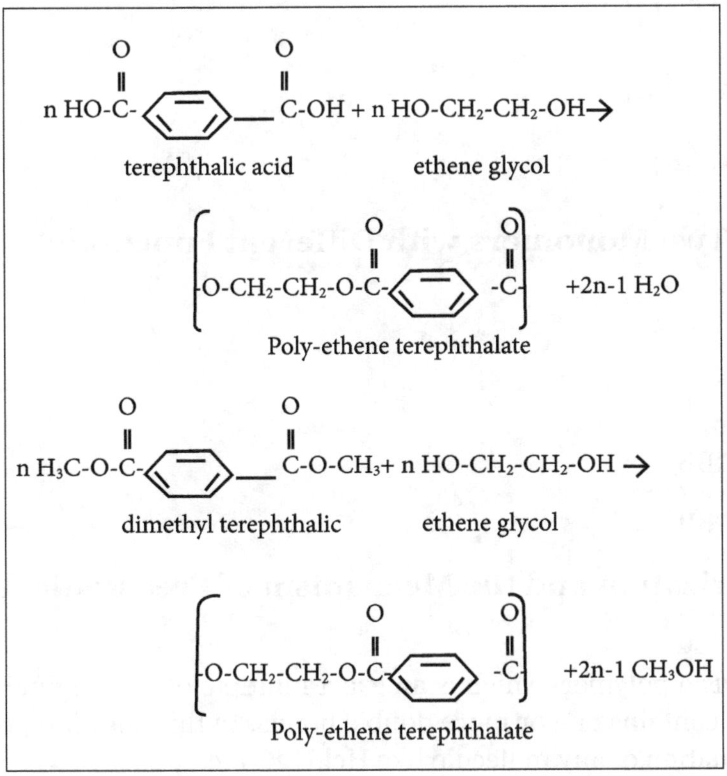

In chain-growth polymerization, a high molecular weight product is produced right from the start, while the monomer quantity deceases slowly with time. In contrast, in step-growth polymerization there is a slow increase in average molecular weight of the product. The molecular weight is usually not as high as in chain-growth polymerization and relatively small amounts of unreacted monomer are present after the start of the reaction.

Chain-growth polymerization generally is fast, irreversible and moderately to highly exothermic.

On the other hand step-growth polymerization is usually slow, equilibrium-limited and isothermal to slightly exothermic.

Polymerization of Bifunctional Monomer (A, B: Two Different Functional Groups)

$$AB + AB \rightarrow ABAB\ (\text{or } AB)_2$$

$$AB + (AB)_2 \rightarrow (AB)_3$$

$$AB + (AB)_3 \rightarrow (AB)_4$$

$$(AB)_2 + AB)_2 \rightarrow (AB)_4$$

$$(AB)_r + (AB)_s \rightarrow (AB)_{r+s}$$

$$(AB)_r + (AB)_s \rightarrow (AB)_{r+s}$$

Polymerization of Two Monomers with Different Functional Groups

AA+BB→AABB

AABB+AA→AABBAA

AABBAA+BB→AABBAABB

AABB+AABB→AABBAABB

Additional Polymerization and the Mechanism of Free Radical Polymerization

It is a reaction that yields a polymer, which is an exact multiple of the original monomeric molecule usually, contains one or more double bonds. In this addition polymerization there is no elimination of any molecule like Hcl,H2O etc.

Example: PVC is produced from vinyl chloride.

$$nCH_2 = \underset{\displaystyle Cl}{\underset{|}{CH}} \xrightarrow{\text{Heat / Pressure}} n..\!-\!\left(CH_2\text{-}\underset{\displaystyle Cl}{\underset{|}{CH}}\right)_n\!-$$

The mechanism of addition polymerization can be explained by one of the following types:

- Free radical mechanism.
- Ionic mechanism.

All the above mechanism occur in three major steps namely:

- Initiation.
- Propagation.
- Termination.

1. Initiation

It is considered to involve two reactions:

i. First reaction involves production of free radicals by homolytic dissociation of an initiator (or catalyst) to yield a pair of free radicals (R*),

$$I \rightarrow 2R^*$$

(Initiator)(Free radicals)

Commonly used Thermal Initiators

Thermal initiator is a substance used to produce free radicals by homolytic dissociation at high temperature,

$$\underset{\text{Acetyl peroxide}}{CH_3COO\text{-}OOCH_3} \xrightarrow{\Delta} \underset{\text{Free radicals}}{2(CH_3COO)(or)2R^{\circ}}$$

ii. Second reaction involves addition of this free radical to the first monomers to produce chain initiating species.

$$\underset{\text{(Free radical)}}{R} + \underset{\text{First monomer}}{CH_2 = \overset{\displaystyle H}{\overset{|}{\underset{\displaystyle Y}{\underset{|}{C}}}}} \longrightarrow \underset{\text{Chain initating species}}{R - CH_2 = \overset{\displaystyle H}{\overset{|}{\underset{\displaystyle Y}{\underset{|}{C}}}}}$$

2. Propagation

It involves the growth of chain initiation species by successive addition of large number of monomers.

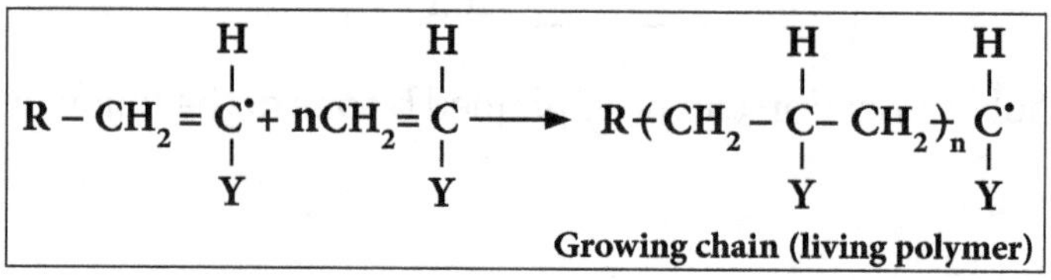

The growing chain of the polymer is known as living polymer.

3. Termination

Termination of the growing chain of polymer may occur either by coupling reaction or disproportionation.

I. Coupling or Combination

It involves coupling of free radical of one chain end to another free radical forming macromolecule so called as dead polymer.

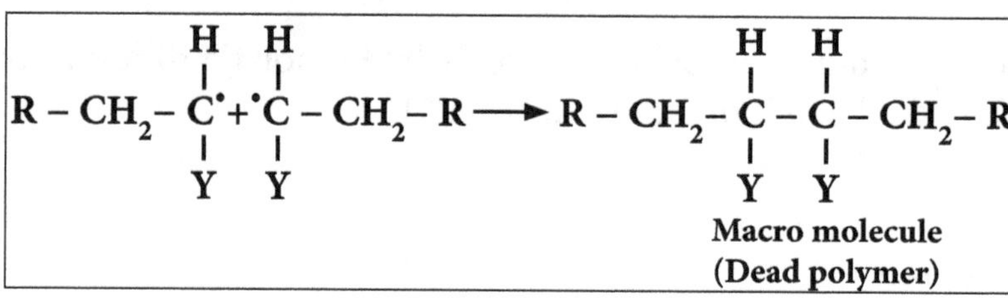

II. Disproportionation

It involves transfer of a hydrogen atom of one radical center to another radical center forming two macro molecules one saturated and another unsaturated.

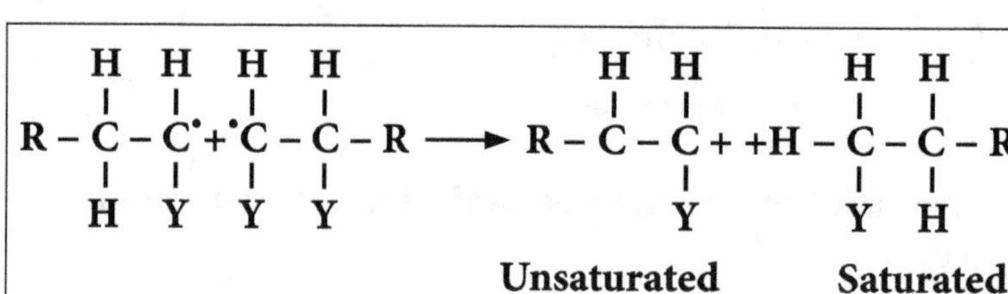

Physical Properties of Polymers

- Strength: It depends upon the magnitude of force of attraction between polymeric chains. Two types (a) Primary or chemical bond and (b) Secondary or intermolecular forces (van der Wall force or hydrogen bonding).

- Deformation: Deformation is the slipping of one chain over the other (on the application of heat or pressure or both) or stretching and recoverance of original shape of the polymeric chains (after the removal of stress).

Types

- Plastic Deformation (Plasticity).
- Elastic Deformation (Elasticity).

3. Chemical Resistance and Solubility

Chemical attack is internal, causing softening, swelling and loss of strength of polymer. The chemical nature of monomeric units and their molecular arrangement determines the chemical resistance of the polymer.

Polymers Polymers having polar groups (-OH, –COOH or Cl) swollen swollen or even dissolved in polar solvents whereas polymers having nonpolar groups ($-CH_3$ or $-C_6H_5$) swollen or even dissolved in non-polar solvents.

Polymers of more aliphatic character are more soluble in aliphatic solvents whereas polymers of more aromatic character are more soluble in aromatic solvents.

The tendency to swell or solubility of polymers decreases with the increase in chain length or molecular weight of polymer. High molecular weight polymers on dissolving yield solutions of high viscosities.

4. Physical state of Polymers

An amorphous state is characterized by a completely random, irregular and dissymmetrical arrangement of polymeric chains. e.g., rubber, thermosetting.

A crystalline state is characterized by a completely regular, symmetrical and ordered arrangement of polymeric chains with uniaxial orientation. e.g., fibers.

A semi-crystalline state consists of crystalline region called crystallites (ordered arrangement of polymeric chains) embedded in an amorphous matrix. e.g., thermoplastic.

3.3.2 Mechanical Properties of Polymers

1. Hardness

The ability of a polymer to resist scratching, abrasion, cutting or penetration. It is measured by its ability to absorb energy under impact loads. Hardness is associated with strength. It is closely associated with material structure, composition and other mechanical properties.

2. Toughness

It is the amount of energy a polymer can absorb before actual fracture or failure takes place. The ability of a polymer to withstand shock and vibrations. It is related related to impact strength strength which is the resistance resistance to breakage under high velocity impact conditions, i.e., resistance to shock loading. It is the ability of a polymer to withstand both plastic and elastic deformation.

3. Stiffness

The resistance of a polymer to elastic deformation, i.e., a polymer which suffers slight deformation under load has a high degree of stiffness. Flexibility (has to do with bending) is the opposite of stiffness.

4. Density

- Relative Density is the mass of the polymer with the mass of equal volume of a specific (reference) substance (water). Density is frequently measured as a quality control parameter.

A specimen with smooth surfaces from crevices and dust, is weighed in the air (W_1) and then in freshly boiled water (W_2),

Then,

$$\rho_{\text{polymer}} = \frac{W_1}{W_1 - W_2} \rho_{\text{water}}$$

5. Tensile Strength

The strength of a polymer is its capacity to withstand destruction under the action of loads. It determines the ability of a polymer to withstand stress without failure. Tensile strength or ultimate strength is the stress corresponding to the maximum load reached before rupturing the polymer,

$$\text{Tensile strength or stress} = \frac{\text{Force or Maximum Load}}{\text{Area of Cross section}}$$

Abrasion Resistance

It is defined as the ability of a polymer to withstand mechanical action (such as scrapping, rubbing or erosion) that tends progressively to remove material from its surface. Abrasion is closely related to the frictional force, the load and true area of contact, an increase in any one of the three results in greater abrasion or wear. Abrasion process also creates oxidation on the surface from the buildup of localized high temperatures.

Resilience

It is the capacity of a polymer to absorb energy elastically Resilience gives capacity of the polymer to bear shocks and vibrations. When a body is loaded, it changes its dimension, dimension and on the removal of the load it regains its original dimensions.

In fact, the polymer behaves perfectly like a spring, so long as it remains loaded, it has stored energy in itself, on removal of the load, the energy stored is given off exactly as in a spring when the load is removed.

Wear and Tear

It occurs when a steady rate of increase in the use of polymers in bearing applications and in situations where there is sliding contact e.g. gears, piston rings, seals, cams, etc. Wear and tear is characterized by fine particles of polymer being removed from the surface or the polymer becomes overheated to the extent where large troughs of melted polymer are removed.

The wear and tear of polymers is extremely complex subjects which depend markedly on the nature of the application and the properties of the material. It is characterized by adhesion and deformation which results results in frictional frictional forces that are not proportional to load but rather to speed. The mechanism of wear and tear is complex the relative rates may change depending on specific circumstance.

3.3.3 Fracture in Polymers

Fracture strengths of polymers are low compared to metals and ceramics ¾ Brittle fracture occurs in thermosetting polymers.

Fracture is initiated at stress concentrators (scratches, notches, etc.). Covalent bonds are severed during fracture.

In thermoplastic polymers, both ductile and brittle fracture are possible. Brittle fracture is favored at lower temperatures, higher strain rates and at stress concentrators.

Brittle to ductile transition often occurs with increasing temperature.

Fracture of Polymers: Crazing

Fracture of glassy thermoplastic polymers often proceeds through crazing. Crazing occurs when localized regions yield, forming an interconnected array of micro voids. Fibrillar bridges of oriented molecular chains form between voids.

At high enough tensile loads, these bridges elongate and break, enabling crack propagation.

Crazing absorbs fracture energy and increases fracture toughness.

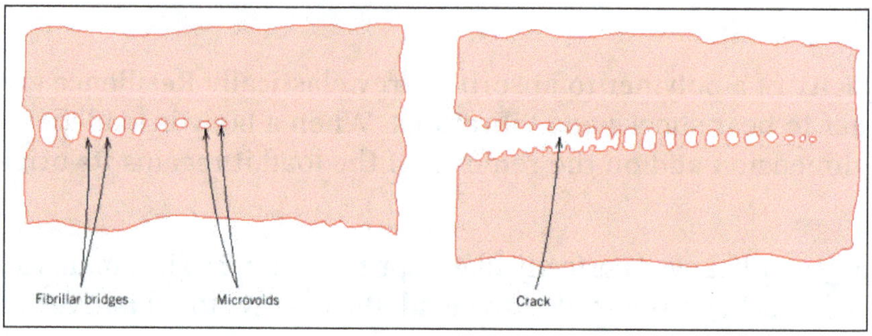

Crazing.

3.4 Rubber: Types and Applications

There are plenty of different rubber types available but still the largest single type used is natural rubber (NR) produced of latex from the tree Hevea Brasiliensis. The synthetic types of polymers mainly manufactured from oil have been developed either to replace or to be used together with NR or to make polymers with properties superior of NR in special areas typically with better high temperature resistance better outdoor resistance or resistance to fuels and oils.

Different types of rubber:

- General purpose elastomers.
- Special purpose elastomers.
- Speciality elastomers.

General purpose elastomers comprise:

- Natural rubber (NR).
- Polyisoprene rubber (IR).
- Styrene-butadiene rubber (SBR).
- Butadiene rubber (BR).

These types have good physical properties good process ability and compatibility are generally economical and are typical polymers used in tyres and mechanical rubber goods with demand for good abrasion resistance and tensile properties. General purpose types constitute the largest volume of polymer used.

Special purpose elastomers comprise:

- Ethylene-propylene rubber (EPM and EPDM).

- Butyl rubber (IIR).
- Chloroprene rubber (CR).
- Acrylonitrile-butadiene rubber or Nitrile rubber (NBR).

They have all unique properties which cannot be matched by the general purpose types and are very important for manufacturing of industrial and automotive rubber products.

The Speciality elastomers are a great number of polymers with very special properties in many cases of great importance for the automotive, aircraft, space and offshore industries. Some of these polymers are:

- Chlorosulfonated Polyethylene (CSM).
- Acrylic Rubber (ACM).
- Silicone Rubber (PMQ/PV/MQ/VMQ).
- Floursilicone Rubber (FPQ).
- Fluoroelastomers (FPM/FFKM/FEPM).
- Urethane Rubber (AU/EU).
- Epichlorohydrine Rubber (CO/ECO/GECO).

Drawbacks of natural rubber:

- Easy ageing.
- Cannot resist strong acid and oil.
- Can resist plant oil.

3.4.1 Thermosetting and Thermoplastics

Thermoplastics

A thermoplastic is a material which becomes soft when heated and hard when cooled. Thermoplastic materials can be cooled and heated several times.

Thermosetting Plastics

Thermosetting plastics cannot be remoulded because all the polymer chains are linked by strong covalent bond (cross links).

Differences between thermoplastics and thermosetting plastics:

Thermoplastic resins	Thermosetting resins
They are formed by addition polymerization.	They are formed by condensation polymerization.
They consist of linear long chain polymer.	They consist of three dimensional network structure.
All the polymer chain are held together by weak Vander Waal's forces.	All the polymer chain are linked by strong covalent bond.
They are weak, soft and less brittle.	They are strong hard and more brittle.
They soften on heating and harden on cooling.	They do not soften on heating.
They can be remoulded.	They cannot be remoulded.
They have low molecular weights.	They have high molecular weight.
They are soluble in organic solvents.	They are insoluble in organic solvents.
Eg: i) Polyethylene ii) Polyvinyl chloride	Eg: i) Bakelite ii) Polyester iii) Urea - formaldehyde

3.4.2 Conducting Polymers

An organic polymer with de localised Π-electron system, having electrical conductance of the same order of a conductor is called conducting polymer:

- This is one of the simplest chain polymers it possesses conjugated double bond.

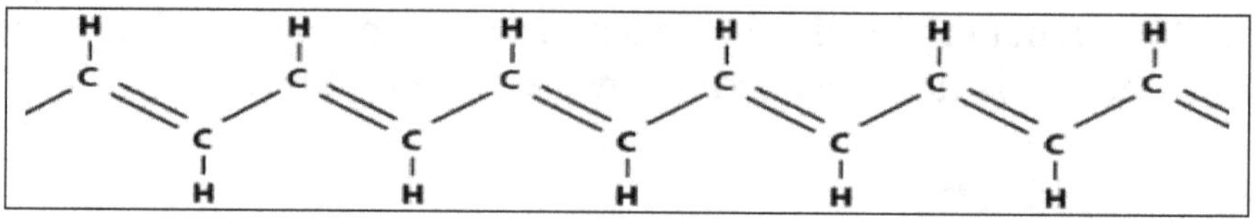

- Conducting polymers are obtained generally by doping (Doping is either the addition of electrons (reduction reaction) or the removal of electrons (oxidation reaction) from an organic polymer with conjugated backbone (consisting of alternating double and single carbon-carbon bonds.

- In a conjugated system, as the electrons are only loosely bound, electron flow may be possible. However as the polymers are covalently bonded the electrons are localized and do not take part in conductivity. Hence the material needs to be doped for electron flow to occur.
- Once doping has occurred, the electrons in the pi-bonds are able to "jump" around the polymer chain. As the electrons are moving along the molecule, electric current occurs.

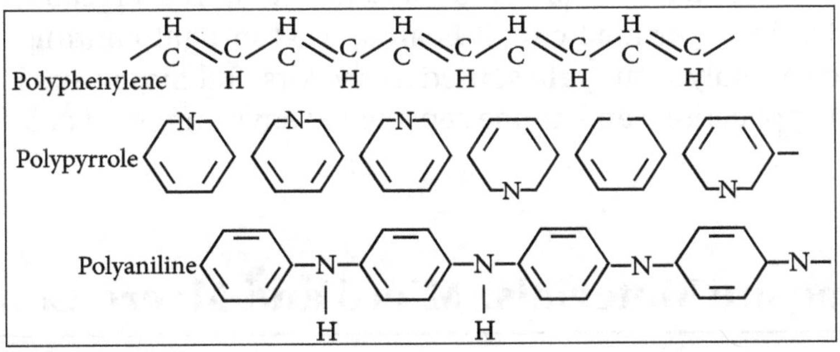

Polyaniline: Distilled 0.2M aniline dissolved in pre cooled 1 M HCI solution and maintained at 50C. Calculated amount of 0.5M $(NH_4)_2S_2O_8$(oxidizing dopant) dissolved in 1M HCI is added slowly to above prepared solution. Since the polymerization reaction is exothermic, to maintain temperature of 0 to 50C the reaction vessel is placed in ice bath. After the complete addition of $(NH_4)2S_2O_8$ the mixture is stirred for about two hours.

Polyaniline finally precipitates in the green form. The dark green precipitate was washed with 1M HCl until the green color disappears. Polyaniline in undoped state is a semiconductor with conductivity 10^{-8} S/cm.

Different oxidation states of polyaniline are:

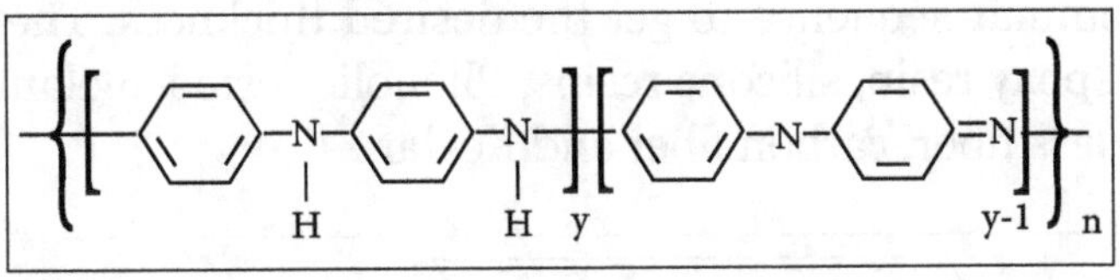

y = 1 Leuco emeraldine

y = 0.5 Emaraldine base

y = 0 Pernigraniline

Emaraldine base consisting of equal proportions of amine (-NH-) and imine (=N-) protonic acid doping, imine sites are protonated by acids to form bipolaron bands.

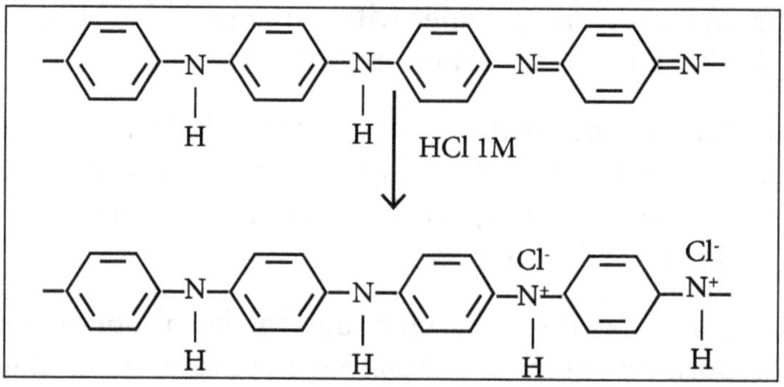

Applications: Used as electrode material for commercial rechargeable batteries, used as conductive tracks on printed circuit boards, used in light emitting diodes, in fuel cells as the electro catalytic materials, used as sensors and biosensors in glucose sensing for medical applications and amine sensing for environmental and food safety applications.

3.5 Composite Materials: Micro and Macro Composites

Combination of two or more structurally different components combine to form a new class of material suitable for structural applications is referred as composite material. When one of the component is a polymer, the resulting composite is called a polymer composite.

Fiber reinforced composites are generally strong and light weight. They are stronger than the steel with less weight and hence they can be used in the automobiles to achieve fuel efficiency and less pollution. They find higher strength per unit weight, low cost of fabrication and resistant to the corrosion.

Polymeric composites are produced by suitably bonding a fiber material with the polymer resin matrix and curing the same under pressure and heat. Alternate layers of resin and fiber are laid in a similar sequence to get the desired thickness. The polymer matrixes generally used are epoxy resin, silicone resins, Phenolic resins, melanine resins, etc. The fibers used as ate-glass fiber, carbon fiber and Kevlar.

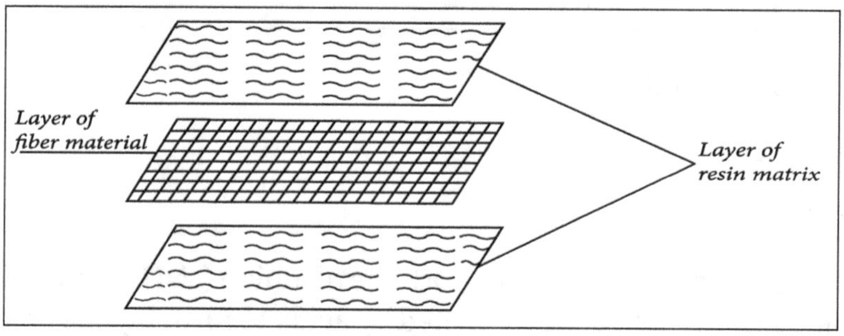

Kevlar (Aramid): It is synthesized from the monomers 1,4-phenylene diamine(para phenylene diamine) and a terephthaloyl chloride through the condensation reaction with the liberation of HCl as a byproduct.

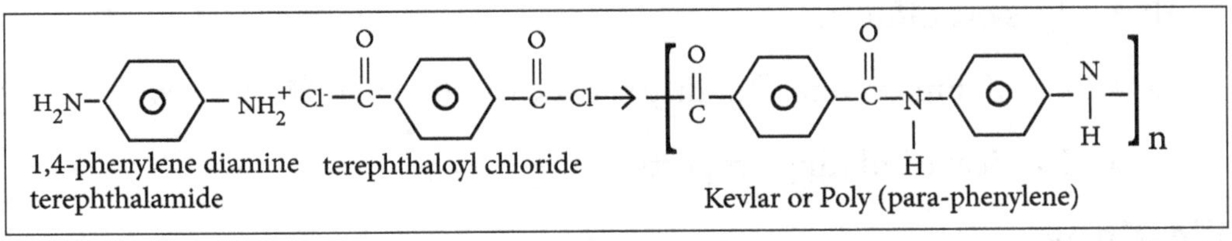

It is high strength due to the presence of inter chain hydrogen bonds formed between the carbonyl group and NH group. Additional strength is achieved from aromatic staking interactions between the adjacent strands. Kevlar is a synthetic fiber stronger than steel.

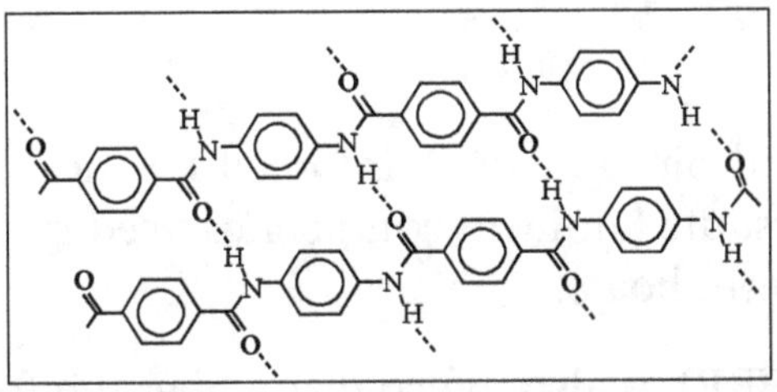

Molecular structure of Kevlar, dashed line indicates hydrogen bonding.

Properties

High chemical inertness, high tensile strength, very low coefficient of thermal expansion, high impact resistance, flame resistance and low weight.

Uses: Used as inner lining for tires to prevent the punctures, used in tennis, table tennis, badminton and squash racquets, hockey sticks, cricket bats, used in personal armor such as helmets, ballistic face masks, bullet proof vests etc. It is often used in the field of cryogenics for its low thermal conductivity and its high strength. Used in the boat hulls, helicopter blades, etc.

3.5.1 Fibre Reinforced Composites

Fiber Reinforced Plastics (FRP) is the generic term for a uniquely versatile family of composites used in everything from the chemical plant to luxury power boats.

An FRP structure typically consists of an unsaturated polyester (UP) resin applied to a mould in combination with the reinforcement, most commonly glass fiber, to form a part that is rigid, highly durable and low in weight.

FRP provides an unrivalled combination of properties:

- Light weight.
- Design freedom.

- High strength-to-weight ratio (kilo-for-kilo it's stronger than steel).
- Chemical resistance.
- High levels of stiffness.
- Retention of dimensional stability across a wide range of temperatures.
- Good electrical insulating properties.

Applications

- Building and Construction.
- Transportation.
- Marine industry.
- Chemical plant and pipes Fiber Reinforced Plastics (FRP) is the generic term for a uniquely versatile family of composites used in everything from chemical plant to luxury power boats.

Fiber reinforced plastic (FRP), is also referred as a fiber reinforced polymer, it is in fact a fiber reinforced plastics composite material constituting a polymer matrix blended with certain reinforcing materials, such as fibers. The fibers are generally basalt, carbon, glass or aramid; in certain cases asbestos, paper or wood can also be used.

An FRP structure typically consists of an unsaturated polyester (UP) resin applied to a mould in combination with reinforcement, most commonly glass fiber, to form a part that is rigid, highly durable and low in weight.

3.5.2 Continuous Fibre Composites and Short Fibre Composites

Composites in which the reinforcing fibres are long, extending over the entire length of the material are known continuous fibre composites. These composites are highly anisotropic, in the sense that their properties vary widely in different directions. The tensile or compressive strength of a continuous fibre composite is very high in the direction parallel to the fibres and quite low in the direction perpendicular to the fibres.

Tensile Stress, Strain Behavior and Longitudinal Loading

Mechanical responses of this type of composite depend on several factors to include the stress–strain behaviors of fiber and matrix phases, the phase volume fractions, and, in addition, the direction in which the stress or load is applied. Furthermore, the properties of a composite having its fibers aligned are highly anisotropic, that is, dependent on the direction in which they are measured.

Let us first consider the stress–strain behavior for the situation wherein the stress is applied along the direction of alignment, the longitudinal direction, which is indicated in Figure (a). To begin, assume the stress versus strain behaviors for fiber and matrix phases that are represented schematically in Figure (a); in this treatment we consider the fiber to be totally brittle and the matrix phase to be reasonably ductile.

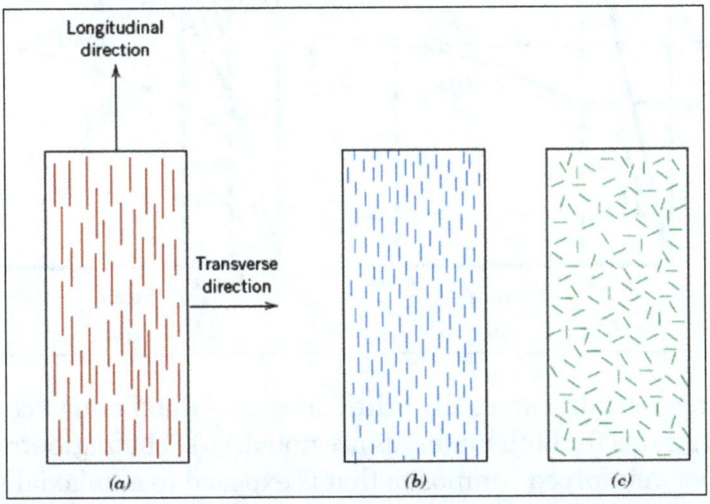

(a) Continuous and aligned, (b) discontinuous and aligned and (c) discontinuous and randomly oriented fiber-reinforced composites.

Also indicate in this figure are fracture strengths in tension for fiber and matrix and respectively and their corresponding fracture strains and furthermore, it is assumed that which is normally the case. A fiber-reinforced composite consisting of these fiber and matrix materials will exhibit the uniaxial stress–strain response illustrated in Figure (b); the fiber and matrix behaviors from Figure (a) are included to provide perspective. In the initial Stage I region, both fibers and matrix deform elastically; normally this portion of the curve is linear. Typically, for a composite of this type, the matrix yields and deforms plastically at figure (b) while the fibers continue to stretch elastically, inasmuch as the tensile strength of the fibers is significantly higher than the yield strength of the matrix. This process constitutes Stage II as noted in the figure; this stage is ordinarily very nearly linear, but of diminished slope relative to Stage I.

Furthermore, in passing from Stage I to Stage II, the proportion of the applied load that is borne by the fibers increases. The onset of composite failure begins as the fibers start to fracture, which corresponds to a strain of approximately as noted in Figure b. Composite failure is not catastrophic for a couple of reasons. First, not all fibers fracture at the same time, since there will always be considerable variations in the fracture strength of brittle fiber materials. In addition, even after fiber failure, the matrix is still intact inasmuch as (Figure a). Thus, these fractured fibers, which are shorter than the original ones, are still embedded within the intact matrix and consequently are capable of sustaining a diminished load as the matrix continues to plastically deform.

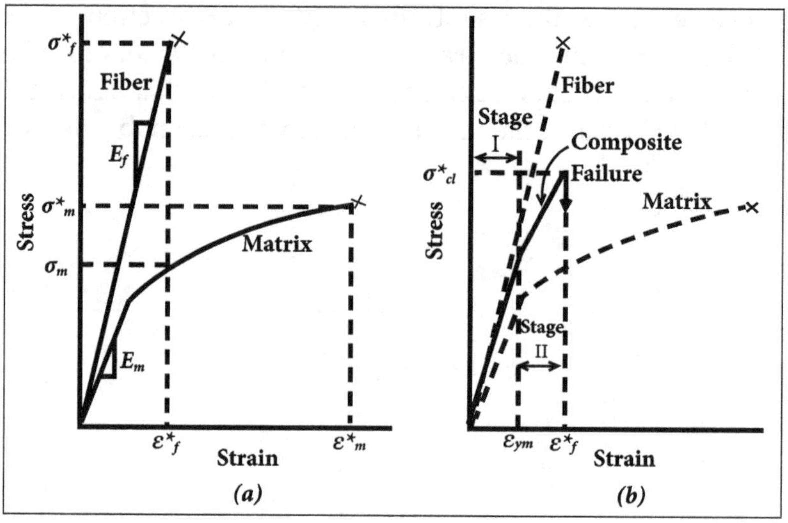

(a) Schematic stress–strain curves for brittle fiber and ductile matrix materials. Fracture stresses and strains for both materials are noted. (b) Schematic stress–strain curve for an aligned fiber-reinforced composite that is exposed to a uniaxial stress applied in the direction of alignment; curves for the fiber and matrix materials shown in part (a) are also superimposed.

Transverse Tensile Strength

The strengths of continuous and unidirectional fibrous composites are highly anisotropic and such composites are normally designed to be loaded along the high strength, longitudinal direction. However, during in-service applications transverse tensile loads may also be present. Under these circumstances, premature failure may result inasmuch as transverse strength is usually extremely low it sometimes lies below the tensile strength of the matrix. Thus, in actual fact, the reinforcing effect of the fibers is a negative one. Typical transverse tensile strengths for three unidirectional composites are contained in Table. Whereas longitudinal strength is dominated by fiber strength, a variety of factors will have a significant influence on the transverse strength; these factors include properties of both the fiber and matrix, the fiber matrix bond strength and the presence of voids. Measures that have been employed to improve the transverse strength of these composites usually involve modifying properties of the matrix.

Material	Longitudinal Tensile Strength (M Pa)	Transverse Tensile Strength (M Pa)
Glass – polyester	700	20

Carbon (high modulus)- epoxy	1000	35
Kevlar™ – epoxy	1200	20

Short Fibre Composites

Cost effective manufacturing often drives the use of discontinuous or "short" reinforcing elements in composites. These materials differ from continuous reinforced composites because the fibers do not extend throughout the entire material. Therefore, load is not directly applied to each reinforcing element.

Stress Transfer

The most commonly quoted theory of stress transfer in discontinuous element composites is the shear-lag analysis. Classically, this element is considered as a discontinuous cylindrical fiber embedded in a continuous matrix. Differences in longitudinal strain in the matrix and adjacent fiber (matrix > fiber) will result in shear stresses at the interface. Tensile load is transferred to the fibers by a shearing mechanism between fibers and matrix. The stresses acting on this body are,

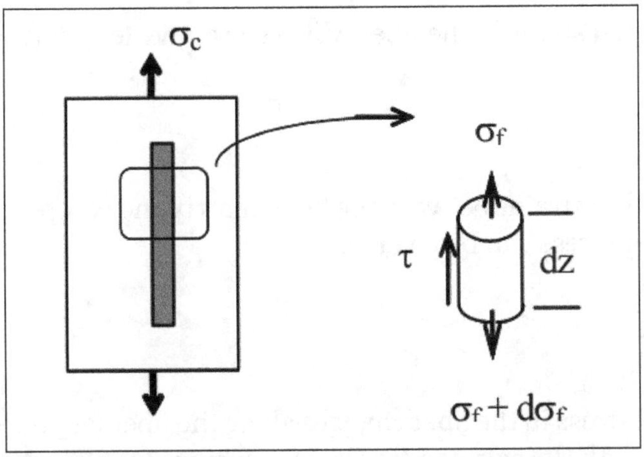

Ignoring stress transfer at the fiber end cross sections and interaction between neighboring fibers, we can calculate normal stress distribution by a simple equilibrium analysis. Consider the force equilibrium on the infinitesimal length, dz,

$$\left(\pi r^2\right)\sigma_f + (2\pi r\,dz)\tau = \left(\pi r^2\right)\left(\sigma_f + d\sigma_f\right)$$

Reducing this relation, we find that the change in fiber stress with length is:

$$\frac{d\sigma_f}{dz} = \frac{2\tau}{r}$$

This relation implies that the change in fiber stress is proportional to the shear stress at the interface between the fiber and matrix. The fiber stress at any point along the fiber is obtained by integrating this equation with respect to z,

$$\sigma_f = \sigma_{fo} + \frac{2}{r}\int_0^z \tau dz$$

The stress at the fiber end, σfo, is often negligible and can be neglected in the formulation,

$$\sigma_f = \frac{2}{r}\int_0^z \tau dz$$

This formulation requires knowledge for the shear stresses along the fiber length. Although these are seldom known, one frequently used assumption is that the matrix material is perfectly rigid and perfectly plastic. In this case, the shear stress would develop instantly at the end of the fiber, reaching a maximum of the yield stress (τy). The normal stresses in the fiber can then be given as:

$$\sigma_f = \frac{2\tau_y z}{r}$$

The maximum normal stress in the fiber will occur a mid-length (z = l/2),

$$\sigma_f^{max} = \frac{\tau_y l}{r}$$

Assuming continuity of strains between the fiber, matrix and composite $(\text{i.e. } \varepsilon_f = \varepsilon_m = \varepsilon_c)$, the maximum fiber stress can be given as,

$$\sigma_f^{max} = \frac{E_f}{E_c}\sigma_c$$

Because the normal stress in the fiber changes along the fiber length, we can assume that a minimum fiber length (lt) exists to transfer the maximum fiber stress. This minimum fiber length can be calculated by combining the previous two equations and rearranging,

$$\frac{l_t}{d} = \frac{\sigma_f^{max}}{2\tau_y} \quad \frac{(E_f / E_c)\sigma_c}{2\tau_y}$$

If we consider that the maximum fiber stress can be equated to that required to fail the fiber (σfu), then we can imagine that extremely short fibers will never develop enough stress to fail. This concept leads to the theory that a critical fiber length (lc) exists for effective reinforcement,

$$\frac{l_c}{d} = \frac{\sigma_{fu}}{2\tau_y}$$

3.6 Polymer Matrix and Metal-Matrix Composites

Polymer Matrix Composite (PMC) is the material consisting of a polymer (resin) matrix combined with a fibrous reinforcing dispersed phase. Polymer Matrix Composites are very popular due to their low cost and simple fabrication methods.

Use of non-reinforced polymers as structure materials is limited by low level of their mechanical properties: tensile strength of one of the strongest polymers - epoxy resin is 20000 psi (140 MPa). In addition to relatively low strength, polymer materials possess low impact resistance.

Reinforcement of polymers by strong fibrous network permits fabrication of Polymer Matrix Composites (PMC) characterized by the following properties:

- High stiffness.
- High Fracture Toughness.
- High tensile strength.
- Good puncture resistance.
- Good abrasion resistance.
- Good corrosion resistance.
- Low cost.

The main disadvantages of Polymer Matrix Composites (PMC) are:

- High coefficient of thermal expansion.
- Low thermal resistance.

Two types of polymers are used as matrix materials for fabrication composites: Thermosets (epoxies, phenolics) and Thermoplastics (High Density Polyethylene (HDPE), Low Density Polyethylene (LDPE), polypropylene, acrylics, nylon).

According to the reinforcement material the following groups of Polymer Matrix Composites (PMC) are used:

- Carbon Fiber Reinforced Polymer Composites.
- Fiber glasses – Glass Fiber Reinforced Polymers.
- Kevlar fiber reinforced polymers.

Metal-Matrix Composites

Metal-matrix composite materials have been so intensely researched over the past years that many new high-strength-to-weight materials have been produced. Most of the materials have been developed for the aerospace industries, but some are being used in other applications such as automobile engines. In general according to reinforcement, the three main types of MMCs arc continuous-fiber, discontinuous-fiber and particulate reinforced.

Continuous-Fiber Reinforced MMCs Continuous filaments provide the greatest improvement in stiffness (tensile modulus) and strength for MMCs. One of the first developed continuous-fiber MMCs was the aluminum alloy matrix boron fiber reinforced system. The boron fiber for this composite is made by chemically vapor depositing boron on a tungsten-wire substrate (figure (a). The Al-B composite is made by hot pressing layers of B fibers between aluminum foils so that the foils deform around the fibers and bond to each other. Figure (b) shows a cross section of a continuous-B-fiber aluminum alloy matrix composite.

With the addition of 51 vol % B. the axial tensile strength of aluminum alloy 6061 was increased from 310 to 1417 MPa. While its tensile modulus was increased from 69 to 231 GPa. Applications for Al-B composites include some of the structural members in the mid fuselage of the space shuttle orbiter.

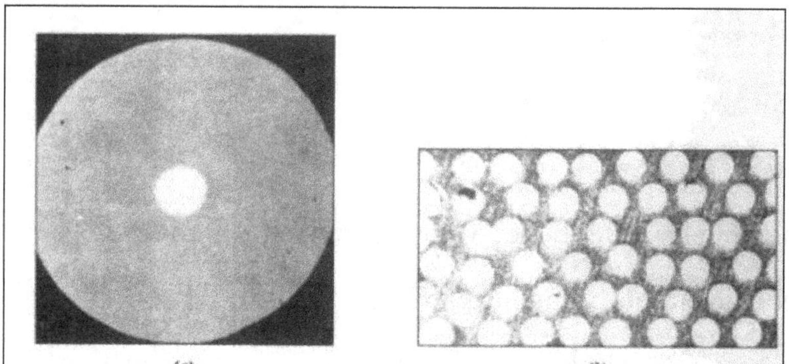

(a) (b)

Discontinuous-Fiber and Particulate Reinforced MMCs Many different kinds of discontinuous and particulate reinforced MMCs have been produced. These materials have the engineering advantage of higher strength, greater stiffness and better dimensional stability than the unreinforced metal alloys. In this brief discussion of MMCs we will focus on aluminum alloy MMCs.

Particulate reinforced MMCs are low-cost aluminum alloy MMCs made by using irregular-shaped particles of alumina and silicon carbide in the range of about 3 to 200μm in diameter. The particulate, which is sometimes given a proprietary coating, can be mixed with the molten aluminum alloy and cast into remelt ingots or extrusion billets for further fabrication. Indicates that the ultimate tensile strength of Al alloy 6061 can be increased from 310 to 496 MPa with a 20 percent SiC addition, while the tensile modulus can be increased from 69 to 103 GPa. Applications for this material include sporting equipment and automobile engine Pans.

Discontinuous-fiber reinforced MMCs are produced mainly by powder metallurgy and melt infiltration processes. In the powder metallurgy process. needlelike silicon carbide whiskers about 1 to 3μm in diameter and 50 to 200pm long are mixed with metal powders, consolidated by hot pressing and then extruded or forged into the desired shape. AI alloy 6061 can be increased from 310 to 480 MPa with a 20 percent SiC whisker addition, while the tensile modulus can be raised from 69 to 115 GPa. Although greater increases in strength and stiffness can be achieved with the whisker additions than with the particulate material, the powder metallurgy and melt infiltration processes arc more costly. Applications for discontinuous-fiber reinforced aluminum alloy MMCs include missile guidance parts and high-performance auto-mobile pistons.

3.6.1 Ceramic-Matrix Composites, Carbon-Carbon Composites, Hybrid Composites

Ceramic-matrix composites have been developed recently with improved mechanical properties such as strength and toughness over the unreinforced ceramic matrix. Again, the three main types according to reinforcement are continuous-fiber, discontinuous-fiber and particulate reinforced. Continuous-Fiber Reinforced CMCs Two kinds of continuous fibers that have been used for CMCs are silicon carbide and aluminum oxide. In one process to make a ceramic-matrix composite, SiC fibers are woven into a mat and then chemical vapor deposition is used to impregnate SiC into the fibrous mat. In another process, SiC fibers are encapsulated by a glass-ceramic material.

(a) (b)

Applications for these materials include heat-exchanger tubes, thermal protection systems and materials for corrosion-erosion environments. Discontinuous (Whisker) and Particulate Reinforced CMCs Ceramic whiskers figure (a And b) can significantly increase the fracture toughness of monolithic ceramics A 20 vol % SiC whisker addition to alumina can increase the fracture toughness of the alumina ceramic from about 4.5 to 8.5 MPa √m. Short-fiber and particulate reinforced ceramic-matrix materials have the advantage of being able to be fabricated by common ceramic processes such as hot isostatic pressing (HIPing).

Ceramic-matrix composites are believed to be toughened by three main mechanisms, all of which result from the reinforcing fibers interfering with crack propagation in the ceramic. These mechanisms are:

- Crack deflection: Upon encountering the reinforcement, the crack is deflected, making its propagating path more meandering. Thus higher stresses are required to propagate the crack.
- Crack bridging: Fibers or whiskers can bridge the crack and help keep the material together, thus increasing the stress level needed to cause further cracking.
- Fiber pullout: The friction caused by fibers or whiskers being pulled out of the cracking matrix absorbs energy.

Carbon-Carbon Composites

In this composite, both the matrix and the reinforcing agent are carbon. The processing technique of this composite is relatively more complex and so they are much more expensive than other composites. They have superior mechanical properties and they are used in special applications like rocket motors, space shuttle structures. High performance automobiles, advanced turbine engines, etc.

Hybrid Composites

Hybrid composites are composites in which two or more types of fibres are used in the same matrix. Most widely used hybrid composites use both glass and carbon fibres in polymer resin matrix. The combination is made so as to achieve improved mechanical properties and maintain low cost. Carbon fibres are much stiffer (higher elastic modulus) and stronger (higher tensile strength) than glass fibres, but they are much more expensive. So the two fibres are effectively used so as to achieve a compromise between strength and cost.

The two types of fibres may be mixed with one another and used in the matrix with all of them aligned parallel or laminates of thin sheets of glass fibre reinforced matrix and carbon fibre reinforced matrix may be stacked alternately. They find applications in sports equipment, automobile components, orthopaedic surgery, etc.

3.7 Ceramics: Types, Structure, Properties and Application

Classification of ceramics materials Ceramics can be classified in diverse ways i.e. there are number of ways to classify the ceramic materials. Most commonly, the ceramics can be classified on the following basis:

- Classification based on applications.

- Classification based on composition.

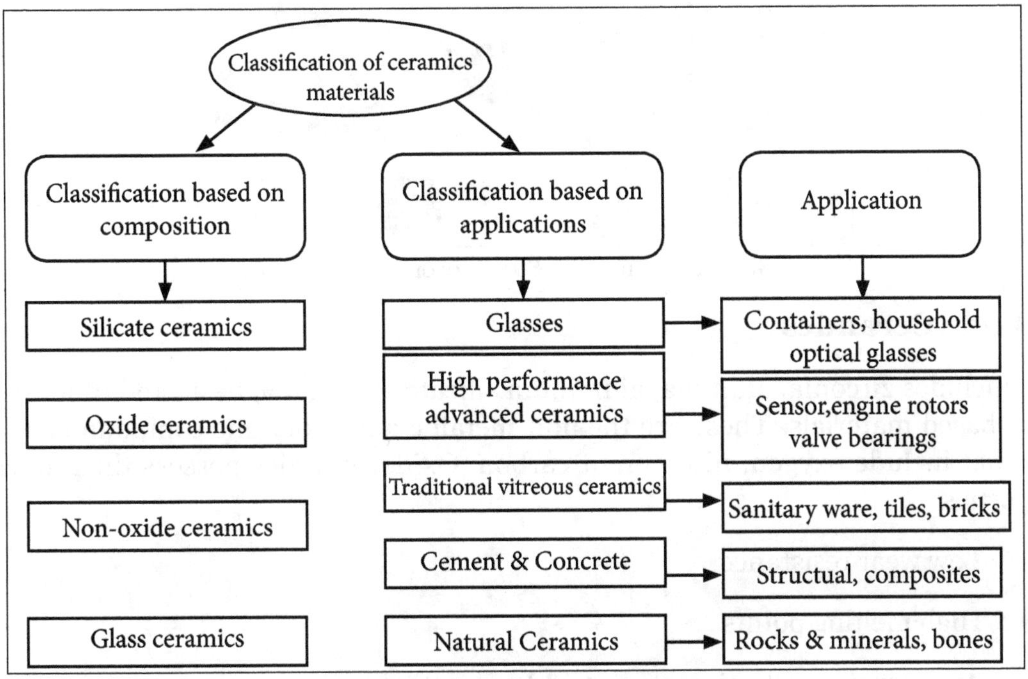

Classification of ceramics materials.

A ceramic material is a non-metallic material, inorganic and is often crystalline. Traditional ceramics are basically clays. The earliest application was in the pottery. Most recently, different types of ceramics used are alumina, silicon carbide etc. The latest advancements are in the bio-ceramics with examples being dental implants and synthetic bones.

A comparative analysis of the ceramics with other engineering materials is shown in table. The purpose of presenting this comparative analysis is to show importance of ceramics among different engineering metals and polymers. This comparison would enable to justify application areas of ceramics.

1. Silicate Ceramics

Silicates are materials generally having composition of silicon and oxygen (figure a). Four large oxygen (o) atoms surround each smaller silicon (Si) atom as shown in figure b. The main types of silicate ceramics are based either on aluminosilicate or on magnesium silicates. Out of these two, the former include clay-based ceramics such as porcelain, earthenware, stoneware, bricks etc.

While the latter consists of a talc-based technical ceramics such as steatite, cordierite and forsterite ceramics. Silicate ceramics are traditionally categorized into coarse or fine and, according to water absorption, into dense (< 2 % for fine and < 6 % for coarse) or porous ceramics (> 2% and > 6 %, respectively).

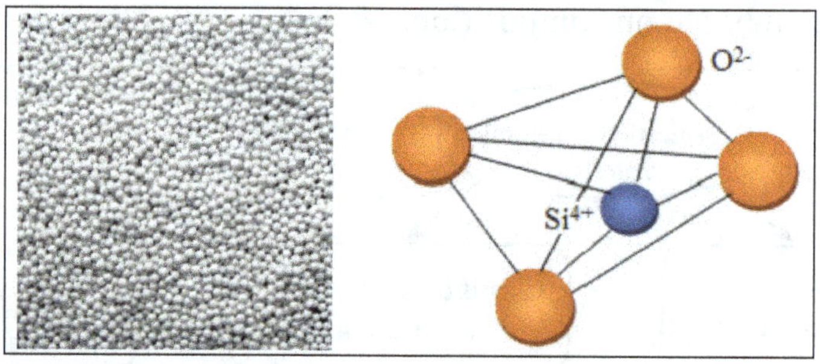

(a) Silicate ceramics; (b) Structure of silicate ceramics.

2. Oxide Ceramics

This includes zirconia, alumina, aluminum silicate, silica, magnesia and other metal oxide based materials. These are the non-metallic and inorganic compounds by nature that include oxygen, nitrogen or carbon. Oxide ceramics possess the following properties:

- Low wear resistance.
- High melting points.
- An extensive collection of electrical properties.

These types of ceramics are available with a variety of special features. For example, glazes and protective coatings seal porosity, improved water or chemical resistance and enhanced joining to metals or other materials.

Oxide ceramics are used in a wide range of applications, which include materials and chemical processing, radio frequency and microwave applications, electrical and high voltage power applications and foundry and metal processing.

The Aluminum oxide (Al_2O_3) is the most important technical oxide ceramic material. This synthetically manufactured material consists of aluminium oxide ranging from 80 % to more than 99 % shown in figure.

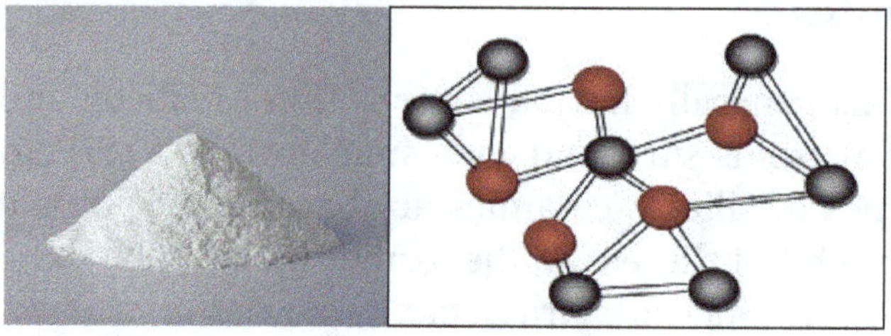

(a) Aluminium oxide (b) Structure of aluminium oxide.

3. Non-Oxide Ceramics

The use of the non-oxide ceramics has enabled extreme wear and corrosion problems to be overcome, even at high temperature and severe thermal shock conditions.

These types of ceramics find its application in different spheres such as pharmaceuticals, oil and gas industry, seals, valves, rotating parts, wear plates, location pins for projection welding, cutting tool tips, abrasive powder blast nozzles, metal forming tooling etc.

4. Glass-Ceramics

These are basically polycrystalline material manufactured through the controlled crystallization of base glass. Glass-ceramic materials share many common characteristics with both glasses and ceramics. Glass-ceramics possess an amorphous phase and more than one crystalline phases. These are produced by a controlled crystallization procedure.

Glass-ceramics holds the processing advantage of glass and has special characteristics of ceramics. Glass-ceramics yield an array of materials with interesting properties like zero porosity, fluorescence, high strength, toughness, low or even negative thermal expansion, opacity, pigmentation, high temperature stability, low dielectric constant, machinability, high chemical durability, biocompatibility, superconductivity, isolation capabilities and high resistivity. These properties can be altered by controlling composition and by controlled heat treatment of the base glass.

Comparison of ceramics with metals and polymers

Property	Ceramic	Metal	Polymer
Density	Low	High	Lowest
Hardness	Highest	Low	Lowest
Ductility	Low	High	High
Wear resistance	High	Low	Low
Corrosion resistance	High	Low	Low
Thermal conductivity	Mostly low	High	Low
Electrical conductivity	Mostly low	High	Low

Applications

- Pottery products, sanitary ware, floor and roof tiles.
- Crucibles, kiln linings, other refractories.
- High end applications such as in ceramic matrix composites, tiles in space shuttle, bullet proof jackets, ball bearing applications, disk brakes, bio-ceramics.

3.8 Other Materials

The material of construction for any application where corrosion is an important factor is selected based on corrosion resistance of the candidate materials to the application environments. Within the economic limits of the application, usually the most corrosion-resistant Material is used. However, even the most corrosion-resistant material is subject to some corrosion, though its rate may be minimal. At times, economics may not permit the use of the most corrosion-resistant material and it may be necessary to use a less costly material with a corrosion rate higher than desirable.

In some applications, an only moderately corrosion-resistant material with some other very desirable properties, such as a high strength-to-weight ratio, high temperature capability or electrical or thermal conductivity, may have to be used because of these properties. When so used, such materials must be protected against corrosion. Because corrosion resistance is a surface-specific property, control measures that alter the surface of the material or the interaction between it and the environment can reduce its corrosion susceptibility.

In doing so, they preserve the physical and mechanical properties of a relatively inexpensive material for an extended life in a given corrosive environment Earlier, when many of the exotic metals, alloys, and non-metallic materials such as high-performance polymers and advanced ceramics, were not commercially available, the only alternative available to the design engineer was to use the commonly available and economic materials with suitable corrosion control measures to reduce corrosion and prolong their lives.

In this, the three broad categories of corrosion control measures are discussed.

Electrochemical Measures

Much corrosion is electrochemical in nature. Anodic or cathodic protection, utilizing an imposed emf, and design measures to minimize galvanic corrosion, can control it.

Passivation Measures

Corrosion can also be reduced by introducing certain minor constituents into the corrosive medium, into the composition of the material, or onto its surface.

Isolation Measure

Corrosion can also be controlled by isolating the material of construction from the corrodent through use of a protective coating or intervening layer of a more corrosion-resistant material, or by suitably modifying the material surface. To the extent corrosion is delayed or reduced, these measures help prolong the life of a product or equipment made of relatively inexpensive materials In most corrosion resistance applications it is desirable to consider such measures to prolong the life of the selected material. While not a substitute for proper selection of material for corrosion control they can make it possible to use an economically viable material to construct a product or equipment, yet min corrosion and pro-long its useful service life. When considering candidate materials for any particular service, use of the relatively less corrosion-resistant and less costly materials, with suitable corrosion control measures, should be considered for comparison with the more costly and corrosion-resistant material.

3.8.1 Nano Phase Materials

Nano particle prepared by precipitation technique:

Nano particles are synthesized by the ppt (precipitation) reaction between the reactants in presence of water soluble inorganic stabilizing agent.

Example: 10gm of sodium hexameter phosphate dissolve in 80 ml of distilled water in 250 ml beaker with straining. Then 10 ml of 1 M sodium sulphate solution strined 1 hr. Precipitation occurs slowly is washed with distilled water and vacuum dried,

$$Ba(NO_3)_2 + Na_2SO_4 \rightarrow BaSO_4 + 2NaNO_3$$

Nanomaterial classification:

- Nano wires.
- Nano rods.
- Nano clusters.
- Nano tubes.

Nano Cluster

These clusters constitute an intermediate state of matter between molecules and bulk materials. These are fine aggregates of atoms or molecules.

Nano Cluster.

Nano Cluster

Nano Cluster constitute an intermediate state of matter between molecules and bulk materials.

They are bound by forces which may be metallic, covalent, ionic hydrogen bond or Vander Waals force in character.

The size of nano-clusters ranges from sub-nanometer to 10 nm in diameter. It has been found that clusters of certain critical size are more stable than others. Nano clusters consisting of upto a couple of hundred atoms. But larger aggregates containing 103 or more atoms are called nano-particles.

Nano–Rods

Nano – rod is a material having an aspect ratio in the range 1 to 20 with short dimension of the material being 10 – 100 mm.

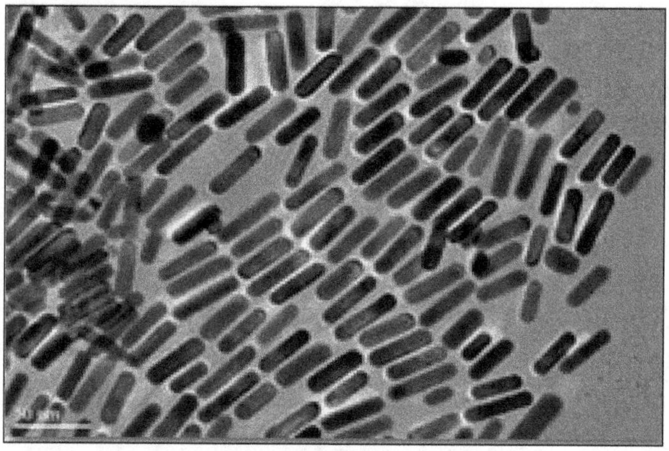

Nanorods.

Synthesis: (Direct Chemical Synthesis)

A Combination of ligands act as a shape control agents and bond to different facets of the nano roads with different strength.

Application

It finds applications display technologies and micro, mechanical switches.

Three uses of nano-rods:

- It finds applications in display technologies.
- It is used to micro mechanical switches.
- Nano rods used for oligo nucleotide detection.

Nano-Tubes (Cnt)

Carbon nanotubes are allotropes of carbon with a nano structure having a length to diameter ratio greater than 1,00,000. When graphite sheets are rolled into a cylinder their edges joined and form carbon nanotubes, i.e., carbon nanotubes are extended tubes of rolled graphite sheets.

Nanotube

- Nano-tubes are one of the most widespread studied and used materials.
- Nano-tubes consists of tiny cylinders of carbon and other materials like boron nitride.
- Nano-tubes of carbon and inorganic compound with structures, comparable to the layered structure of graphite have been prepared. Studies on carbon nano-tubes are quite extensive.

The nano materials and the types of carbon nano tubes and applications:

- Nano materials are die materials having components with size less than 100m at least in one dimension.

Types

Depending upon the way in which graphic sheet rolled, two types of CNTs are formed:

- Single-walled nanotubes (SWNTs).
- Multi-walled nanotubes (MWNTs).

Explanation

Single-Walled Nanotubes (Swnts)

- SWNTs consist of one tube of graphite.
- It is one-atom thick having a diameter of 2nm and a length of 100 μm.
- SWNTs are very important because they exhibit important electrical properties.
- It is an excellent conductor.
- Three kinds of nanotubes are resulted based on the orientation of the hexagon lattice.
    - Arm-chair structure: The lines of hexagons are parallel to the axis of the nanotube.

- Zigzag structure: The lines of carbon bonds are down the center.
- Chiral nanotubes: It exhibits twist or spiral around the nanotubes.

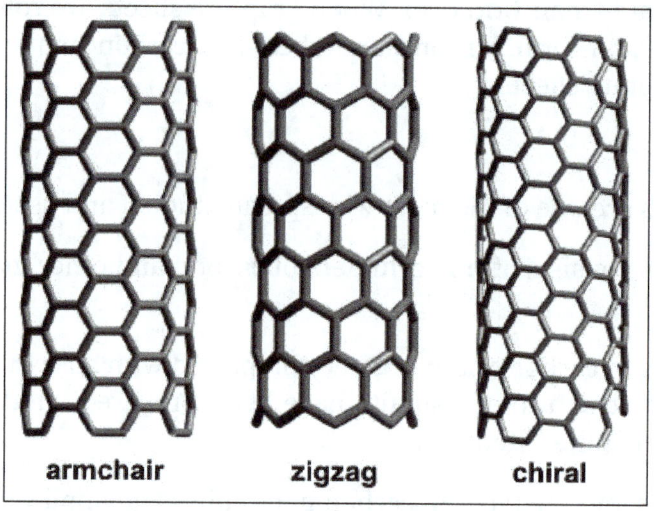

Kinds of nanotubes.

- It has been confirmed that arm-chair carbon nanotubes are metallic while zig-zag and chiral nanotubes are semiconducting.

Multi-Wall Nanotubes (Mwnts)

- MWNTS (nested nanotubes) consist of multiple layers of graphite rolled in on themselves form a tube shape.
- It exhibits both metallic and semi conducting properties.

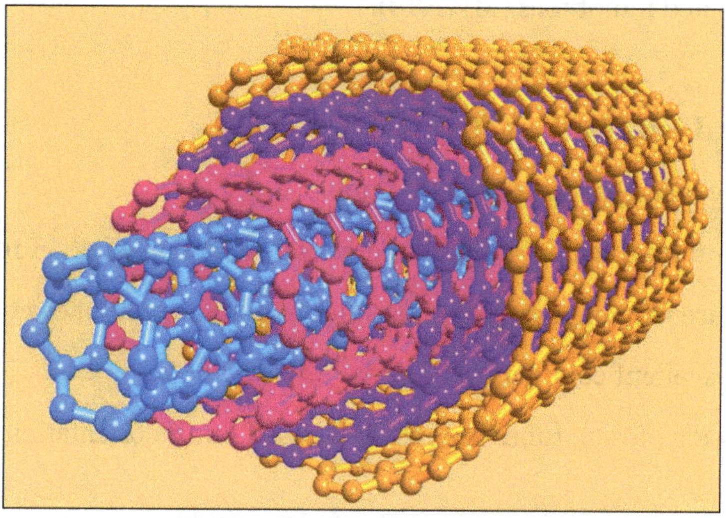

Multi-wall nanotubes (MWNTS).

- It is used for storing fuels such as hydrogen and methane.

Application

- It is used in composites.
- CNTs are used effectively inside the body for drug delivery.
- It is used in battery technology and industries as catalyst.
- It is also used as light weight shielding materials for protecting electronic equipments.

Nano Wire

Nano wire is a nanostructure, with the diameter of the order of a nanometer (10–9 meters). It has length to width ratio greater than 20 nm. Nanowires are also referred to as quantum wires.

Example:

- Nano-wire of metals like Au, Ni, Pt.
- Nano-wire of semiconductor Inp, Si, GaN.
- Nano–wire of insulation: SiO2, TiO2.
- Molecular nanowire: DNA.

Nano wire synthesis by (Template-assisted synthesis).

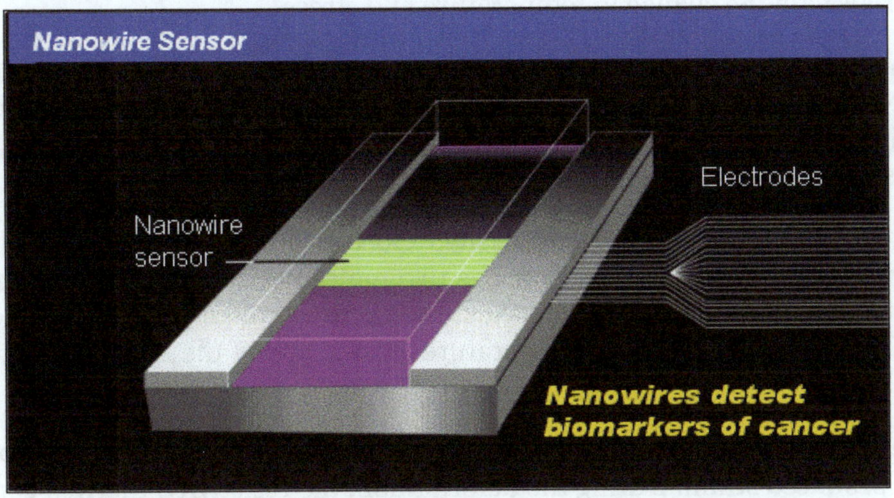

Nano Wire.

3.8.2 Shape Memory Alloy

The ability of the metallic alloys to retain to their original shape when heating or cooling is called as Shape Memory Alloys (SMA). These metallic alloys exhibits the plastic nature when they are cooled to very low temperature and they return to their original nature when they are again heated. This effect is known as Shape Memory Effect.

It is also called as smart materials or intelligent materials or Active materials.

There are two types of shape memory alloys:

- One way shape memory: It returns to its memory only when heating.
- Two way shape memory: It returns to its memory on both heating and Cooling.

Classification

- Piezo electric SMA materials.
- Electrostrictive SMA materials.
- Magnetostrictive SMA materials.
- Thermo elastic SMA materials.

Examples: Ni-Ti (Nickel – Titanium), Cu Zn A1, Cu A1 Ni, Au – Cd, Ni-Mn-Ga and Fe based alloys.

Working Principle of SMA

The shape memory effect occurs in alloys due to change in the crystalline structure of the materials with the change in temperature and stress.

The shape memory effect occurs between two temperature states known as Martensite and Austenite. The Martensite structure is a low temperature phase and is relatively soft and it has platelet structure. The Austenite is a high temperature phase and is hard and it has needle like structure.

Martensite is the relatively soft and easily deformed phase of shape memory alloys which exists at lower temperatures. It has two molecular structures namely, twinned Martensite and deformed Martensite. Austenite is the stronger phase of shape memory alloys which occurs at higher temperatures, the shape of the Austenite structure is cubic.

When we apply a constant load on the shape memory alloy and cool it, its shape changes due to produced strain. During the deformation, the resistivity, thermal conductivity, Young's modulus and yield strength are decreased by more than 40%.

Twinned Martensite state alloy becomes deformed Martensite when it is loaded. The deformed Martensite becomes Austenite when it is heated, the Austenite transformed to original twinned Martensite state when it is cooled.

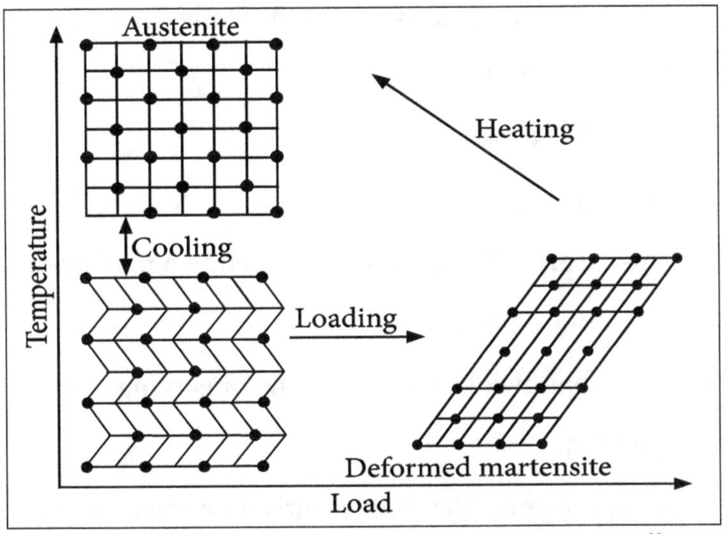

Material crystalline arrangement during shape memory effect.

Pseudo Elasticity

When the metallic material is cooled from a temperature T to a lower temperature TC, it deforms and changes its shape. On reheating the material to temperature (T) the shape change is received so that the material returns to its original state. This effect is known as pseudo elasticity or thermo-elastic property.

Super Elasticity

Super elasticity is a property of SMA. When a material is deformed at a temperature slightly greater than its transformation temperature super elasticity property appears (Rubber like property).

Properties Of Ni – Ti Alloy

Ni – Ti is a compound of Nickel and Titanium and it finds many applications in the field of engineering due to the following properties:

- It has greater shape memory strain.
- It has more thermal stability and excellent corrosion resistance.
- It has higher ductility and more stable transformation temperatures.
- It has better bio-compatibility and it can be electrically heated.

Advantages of SMA's

- They have good bio-compatibility.
- They have simplicity, compactness and high safety mechanism.

- They have good mechanical properties and strong corrosion-resistance.
- They have high power and weight ratio.

Disadvantages of SMA's

- They have poor fatigue properties.
- They are expensive and difficult to be prepared in a machine.
- They have low energy efficiency.
- They have limited bandwidth due to heating or cooling.

Applications of SMA's

- Eye glass frames: We know that the recently manufactured eye glass frames can be bent back and forth and can retain its original shape within fraction of time.
- Toys: We might have seen toys such as butterflies, snakes etc., which are movable and flexible.
- Helicopter blades: The life time of helicopter blades depends on vibrations and their return to its original shape. Hence shape memory alloys are used in helicopter blades.
- Coffee Valves: Used to release the hot milk and the ingredients at a certain temperature.

Medical Applications of SMA's

- It is used as Micro – Surgical instruments.
- It is used as dental arch wires.
- It is used as flow control devices.
- It is used as ortho – dentil implants.
- It is used for repairing of bones.
- They are used to correct the irregularities in teeth.
- Engineering Applications of SMA's.
- It is used as a thermostat valve in cooling system.
- It is used as a sealing plug for high pressure.
- It is used as a fire safety valve.

- It is used for cryolite hydraulic pipe couplings.
- It is used for eye glass frame, toys, liquid safety valve.
- It is used to make microsurgical instruments, orthopedic implants.
- It is used as blood clot filter and for fracture pulling.
- It is used to make antenna wires in cell phones.
- It can be used as circuit edge connector.

3.8.3 Smart Materials

Smart structures arc a new emerging materials system that combines contemporary materials science with information science. The smart system is composed of sensing, processing, actuating, feedback, self-diagnosing, and self-recovering sub-systems. The system uses the functional properties of advanced materials to achieve high performances with capabilities of recognition, discrimination, and ad-justification in response to a change of its environment. Each component of this system must have functionality and the entire system is integrated to perform a self-controlled smart action, similar to a living creature who can "think", judge and act. A smart system can be considered as a design philosophy that emphasizes predicative, adaptivity, and repetition.

A smart system or smart structure is defined to be a non-biological physical structure with:

- A definite purpose.
- Means and imperative to achieve that purpose.
- A biological pattern of functioning.

Smart materials are a subset of the smart system. i.e.. smart structures at the microscopic or macroscopic scales. A smart system is a non-biological structure, which means that the system functions as a biological system rather than a Turing machine. Such a material will generally include at least one structural element, some means of sensing the environment or its own state, and some type of processing and adaptive control algorithm.

Four of the most widely used smart materials nowadays are piezoelectric $Pb(Z_r.T_i)O_3$ (MT), Magnetostriction $(Tb.Dy)Fe_2$, Electrostatics $Pb(Mg.Nb)O_3$ and the shape memory alloy N_iT_i.

Pb(Zr,Ti)O3 is a ferroelectric ceramics which is at a mediate temperature range and becomes ferroelectric on cooling through the Curie temperature (Tc).

At room temperature (RT), it is poised on a rhombohedral-tetragonal phase boundary that enhances the piezoelectric coefficients. Terfeno, $(Tb.Dy)Fe_2$, experiences a rhombohedral-tetragonal transition at RT, which enhances its magneto-striction coefficient. $Pb(Mg.Nb)O_3$ and NiTi are cubic at high temperatures and on annealing, it transforms to a partially ordered state.

On further cooling, $Pb(M_g,N_b)O_3$ passes through a diffuse phase transformation at RT, forming a mixture of ordered and disordered phase domains that exhibits large dielectric and electro astrictive coefficients under certain conditions. Just below RT, it transforms to a ferroelectric rhombohedral phase. The partially ordered shape memory alloy NiTi undergoes an austenitic (bcc) to martensitic (monoclinic) phase change just above RT. It is easily deformed in the martensitic state but recovers its original shape when reheated to austenite.

The structural and phase transformation similarities of these four actuator materials are remarkable and could be a key in developing new smart materials.

Permissions

All chapters in this book are published with permission under the Creative Commons Attribution Share Alike License or equivalent. Every chapter published in this book has been scrutinized by our experts. Their significance has been extensively debated. The topics covered herein carry significant information for a comprehensive understanding. They may even be implemented as practical applications or may be referred to as a beginning point for further studies.

We would like to thank the editorial team for lending their expertise to make the book truly unique. They have played a crucial role in the development of this book. Without their invaluable contributions this book wouldn't have been possible. They have made vital efforts to compile up to date information on the varied aspects of this subject to make this book a valuable addition to the collection of many professionals and students.

This book was conceptualized with the vision of imparting up-to-date and integrated information in this field. To ensure the same, a matchless editorial board was set up. Every individual on the board went through rigorous rounds of assessment to prove their worth. After which they invested a large part of their time researching and compiling the most relevant data for our readers.

The editorial board has been involved in producing this book since its inception. They have spent rigorous hours researching and exploring the diverse topics which have resulted in the successful publishing of this book. They have passed on their knowledge of decades through this book. To expedite this challenging task, the publisher supported the team at every step. A small team of assistant editors was also appointed to further simplify the editing procedure and attain best results for the readers.

Apart from the editorial board, the designing team has also invested a significant amount of their time in understanding the subject and creating the most relevant covers. They scrutinized every image to scout for the most suitable representation of the subject and create an appropriate cover for the book.

The publishing team has been an ardent support to the editorial, designing and production team. Their endless efforts to recruit the best for this project, has resulted in the accomplishment of this book. They are a veteran in the field of academics and their pool of knowledge is as vast as their experience in printing. Their expertise and guidance has proved useful at every step. Their uncompromising quality standards have made this book an exceptional effort. Their encouragement from time to time has been an inspiration for everyone.

The publisher and the editorial board hope that this book will prove to be a valuable piece of knowledge for students, practitioners and scholars across the globe.

Index

www.ingramcontent.com/pod-product-compliance
Lightning Source LLC
LaVergne TN
LVHW080804220926
848485LV00011B/88